*Mössbauer spectroscopy*

# Mössbauer spectroscopy

*EDITED BY*

## DOMINIC P.E.DICKSON
*Department of Physics, University of Liverpool*

## FRANK J.BERRY
*Department of Chemistry, University of Birmingham*

The right of the
University of Cambridge
to print and sell
all manner of books
was granted by
Henry VIII in 1534.
The University has printed
and published continuously
since 1584.

CAMBRIDGE UNIVERSITY PRESS

*Cambridge*

*London   New York   New Rochelle*

*Melbourne   Sydney*

CAMBRIDGE UNIVERSITY PRESS
Cambridge, New York, Melbourne, Madrid, Cape Town, Singapore, São Paulo

Cambridge University Press
The Edinburgh Building, Cambridge CB2 2RU, UK

Published in the United States of America by Cambridge University Press, New York

www.cambridge.org
Information on this title: www.cambridge.org/9780521261012

First published 1986
This digitally printed first paperback version 2005

*A catalogue record for this publication is available from the British Library*

*Library of Congress Cataloguing in Publication data*

Main entry under title:
Mossbauer spectroscopy.

Bibliography : p.
Includes index.
1. Mossbauer spectroscopy. I. Dickson, Dominic P. E.
II. Berry, Frank J., 1947–
QC491.M6135    1986      537.5.′352      85-30905

ISBN-13 978-0-521-26101-2 hardback
ISBN-10 0-521-26101-5 hardback

ISBN-13 978-0-521-01810-4 paperback
ISBN-10 0-521-01810-2 paperback

# Contents

# *Contributors*

E. R. Bauminger
Racah Institute of Physics, The Hebrew University of Jerusalem,
Israel.

F. J. Berry
Department of Chemistry, University of Birmingham, U.K.

S. Dattagupta
Department of Physics, University of Hyderabad, India.

D. P. E. Dickson
Department of Physics, University of Liverpool, U.K.

C. E. Johnson
Department of Physics, University of Liverpool, U.K.

G. J. Long
Department of Chemistry, University of Missouri-Rolla, U.S.A.

I. Nowik
Racah Institute of Physics, The Hebrew University of Jerusalem,
Israel.

R. V. Parish
Department of Chemistry, University of Manchester Institute of
Science and Technology, U.K.

M. F. Thomas
Department of Physics, University of Liverpool, U.K.

# *Preface*

Many techniques have developed in recent years which allow experimental scientists to extend their studies in new directions. An awareness of the scope, strengths, potential, and limitations of a particular technique is essential if it is to be utilised to the full. The aim of this book is to review the unique contribution that Mössbauer spectroscopy can make to the study of the bonding, structural, magnetic, time-dependent, and dynamical properties of various systems. Particular emphasis is given to the types of information which Mössbauer spectroscopy provides, as well as to a comparison with the information in these areas which can be obtained using other techniques.

In order to achieve the objectives outlined above, each of the main chapters relates to one aspect of the information which can be obtained by Mössbauer spectroscopy, and these chapters have been written by scientists with considerable experience in these different areas. We are particularly indebted to the authors of the individual chapters for embracing our overall philosophy for this book and the need to produce a coherent whole, while still reflecting their personal enthusiasm for their own particular area of interests, as well as the differences inherent in these different aspects.

One of the striking features of the growth of Mössbauer spectroscopy, following the discovery by Rudolf Mössbauer in 1957 of the effect which bears his name, was the rapidity with which the potential of the new technique was recognised and developed. This rapid growth owes much to a pioneering generation of Mössbauer spectroscopists who saw the potential and moved into this exciting new field from other areas of Physics and Chemistry. As members of a later wave of Mössbauer spectroscopists, we would like to express our thanks to these pioneers for establishing such a fascinating field for those who follow them.

We wish to dedicate this book to these early workers in this field; in particular to Charles Johnson and Alfred Maddock, who introduced us to the technique, and also to Solly Cohen and Shimon Ofer, two of the pioneers who died during the writing of this book, and who will be sadly missed by the Mössbauer community.

Liverpool and Birmingham                    Dominic P. E. Dickson
1985                                         Frank J. Berry

# 1

## Principles of Mössbauer spectroscopy

DOMINIC P. E. DICKSON &
FRANK J. BERRY

### 1    Introduction

Mössbauer spectroscopy has now established itself as a technique which can provide valuable information in many areas of science. This book sets out to consider the unique contribution of Mössbauer spectroscopy in terms of the specific information it can give to the study of the bonding, structural, magnetic, time-dependent and dynamical properties of various systems. Accordingly, each of the following chapters is devoted to a detailed consideration of one fundamental aspect of the total information that can be provided by Mössbauer spectroscopy. It is intended that by adopting this approach the book will provide an overview and appraisal of the technique that will be of value both to the experienced practitioner who has perhaps worked in one particular area of the technique, and also to the newcomer or student who needs to see the technique in its proper context. It is also hoped that those working in other areas of science may find this book a helpful guide in relating information obtained from Mössbauer spectroscopy with that derived from other techniques and also in evaluating the potential value of Mössbauer spectroscopy for providing information relevant to their particular problems and systems. The importance of Mössbauer spectroscopy as a means by which a variety of problems in a diversity of scientific disciplines may be examined is already well covered in a number of texts (e.g. Cohen, 1976 & 1980; Gonser, 1975 & 1981; Thosar, Srivastava, Iyengar & Bhargava, 1983; Long, 1984) and no attempt is made here to provide coverage of all possible areas of application.

This first chapter provides an introduction to Mössbauer spectroscopy and a background for the subsequent chapters. Hence the next few pages contain a general discussion of the Mössbauer effect, the conditions necessary for its observation, and the way in which Mössbauer spectroscopy has been developed as an experimental technique. The various factors determining the Mössbauer spectrum are then treated in more detail since these relate directly to the information which can be obtained from Mössbauer spectroscopy and hence to the material contained in the remainder of the book. Finally the various areas of application of Mössbauer spectroscopy are briefly considered to give an indication of the ways in which this information can be used.

The nomenclature, symbols and conventions used in this book correspond with those normally used in the field as exemplified in a number of texts which also contain more extensive treatments of the theory and practice of Mössbauer spectroscopy and to which the reader requiring this information is referred (e.g. Gibb, 1977; Gonser, 1975; Greenwood & Gibb, 1971; May, 1971; Wertheim, 1964).

## 2     The Mössbauer effect

For many years it was recognised that the gamma rays emitted when radioactive nuclei in excited states decay might excite other stable nuclei of the same isotope, thereby giving rise to nuclear resonant absorption and fluorescence. However, initial attempts to detect these resonant processes were unsuccessful mainly because of the nuclear recoil, which accompanies both the emission and absorption of the gamma ray by a free nucleus. As a result of this nuclear recoil the energy of the emitted gamma ray is less than the energy difference between the two nuclear levels, while if resonant absorption is to occur the energy of the incoming gamma ray needs to be greater than this energy difference. Thus for free nuclei the recoil energy prevents resonant absorption of gamma rays under normal circumstances.

In 1957 Mössbauer discovered that a nucleus in a solid can sometimes emit and absorb gamma rays without recoil (Mössbauer, 1958); because when it is in a solid matrix the nucleus is no longer isolated, but is fixed within the lattice. In this situation the recoil energy may be less than the lowest quantised lattice vibrational energy and consequently the gamma ray may be emitted without any loss of energy due to the recoil of the nucleus. Since the probability of such a recoil-free event depends on the energy of the nuclear gamma ray the Mössbauer effect is restricted to certain isotopes with low-lying excited states. This probability also depends on the temperature and on the vibrational properties of the solid in which the Mössbauer nucleus is situated. The Mössbauer gamma ray has a finite spread in energy, which is quantified by a linewidth, defined as the full width at half maximum intensity and related to the lifetime of the nuclear excited state. The ease with which the Mössbauer effect is observed for a particular isotope is strongly dependent on the value of this gamma-ray linewidth. This places further restrictions on the isotopes for which Mössbauer spectroscopy is of practical use. In general, the Mössbauer effect is optimised for low-energy gamma rays associated with nuclei strongly bound in a crystal lattice at low temperatures.

The first example of the recoil-free resonant absorption of gamma rays was observed by Mössbauer using $^{191}$Ir (Mössbauer, 1958) and the Mössbauer effect has now been detected in over one hundred isotopes. However, most of the practical applications of the effect in the technique known as Mössbauer spectroscopy have been with a much smaller number of isotopes in which the conditions for observing the Mössbauer effect and obtaining useful information from it are particularly favourable.

Figure 1.1 shows a periodic table of the elements indicating those which have Mössbauer isotopes as well as the ease with which the Mössbauer effect can be observed. As $^{57}$Fe has the most advantageous combination of properties for Mössbauer spectroscopy the great majority of studies continue to involve this isotope and the technique has therefore been widely used for the investigation of iron-containing systems. Since $^{57}$Fe Mössbauer spectroscopy provides good examples of most of the different types of information that can be obtained from this technique, work with this isotope will often be used to illustrate them in this book. However, it should not be forgotten that similar and equally valuable information can also be very effectively obtained with the other isotopes. The detailed

Fig. 1.1. A periodic table of elements with Mössbauer isotopes. The elements with easier to use and more extensively studied isotopes are shown shaded, while those with more difficult to use and less extensively studied isotopes are shown hatched. These two categories were distinguished by whether more or less than one hundred papers were published on work with these Mössbauer isotopes between 1953 and 1982 (Long, 1984). It should be noted that in this period 10 000 papers were published on studies involving the most extensively used isotope $^{57}$Fe.

| H  |    |    |    |    |    |    |    |    |    |    |    |    |    |    |    |    | He |
|----|----|----|----|----|----|----|----|----|----|----|----|----|----|----|----|----|----|
| Li | Be |    |    |    |    |    |    |    |    |    |    | B  | C  | N  | O  | F  | Ne |
| Na | Mg |    |    |    |    |    |    |    |    |    |    | Al | Si | P  | S  | Cl | Ar |
| K  | Ca | Sc | Ti | V  | Cr | Mn | Fe | Co | Ni | Cu | Zn | Ga | Ge | As | Se | Br | Kr |
| Rb | Sr | Y  | Zr | Nb | Mo | Tc | Ru | Rh | Pd | Ag | Cd | In | Sn | Sb | Te | I  | Xe |
| Cs | Ba | La | Hf | Ta | W  | Re | Os | Ir | Pt | Au | Hg | Tl | Pb | Bi | Po | At | Rn |
| Fr | Ra | Ac |    |    |    |    |    |    |    |    |    |    |    |    |    |    |    |

| Ce | Pr | Nd | Pm | Sm | Eu | Gd | Tb | Dy | Ho | Er | Tm | Yb | Lu |
|----|----|----|----|----|----|----|----|----|----|----|----|----|----|
| Th | Pa | U  | Np | Pu | Am | Cm | Bk | Cf | Es | Fm | Md | No | Lr |

characteristics of the various Mössbauer isotopes and the problems and advantages associated with using them are discussed in a number of the basic texts on Mössbauer spectroscopy.

## 3    Mössbauer spectroscopy

The energy levels of a nucleus situated in an atom and in a solid are modified by the environment of the nucleus. Mössbauer spectroscopy is a technique which enables these energy levels to be investigated by measuring the energy dependence of the resonant absorption of Mössbauer gamma rays by nuclei. This is possible since the recoil-free processes arising from the Mössbauer effect lead to the resonant absorption of gamma rays with an extremely precise energy. This enables the very small energy changes resulting from the hyperfine interactions between the nucleus and its surrounding electrons to be investigated and therefore Mössbauer spectroscopy provides a means of using the nucleus as a probe of its environment.

The most usual experimental arrangement for Mössbauer spectroscopy involves a radioactive source containing the Mössbauer isotope in an excited state and an absorber consisting of the material to be investigated which contains this same isotope in its ground state. For example, the source for $^{57}$Fe Mössbauer spectroscopy is normally radioactive $^{57}$Co which undergoes a spontaneous electron capture transition to give a metastable state of $^{57}$Fe which in turn decays to the ground state via a gamma ray cascade which includes the 14.4 keV Mössbauer gamma ray (Figure 1.2).

The radioactive source nuclei are usually embedded in a matrix which provides the necessary solid environment as well as giving the simplest possible hyperfine interaction between these nuclei and their environment. In the normal transmission experiment the gamma rays emitted by the source pass through the absorber, where they may be partially absorbed, and they then pass to a suitable detector. In order to investigate the energy levels of the Mössbauer nucleus in the absorber it is necessary to modify the energy of the gamma rays emitted by the source so that they can have the correct energy for resonant absorption. This is usually accomplished by moving the source relative to a stationary absorber, and hence giving the gamma rays an energy shift as a result of the first-order relativistic Doppler effect. The motion of the source is normally oscillatory in order to provide an energy scan.

The typical Mössbauer spectroscopy experiment described above is depicted in Figure 1.3. Resonant absorption occurs when the energy of the gamma ray just matches the nuclear transition energy for a Mössbauer nucleus in the absorber. The resulting Mössbauer spectrum consists of a plot of gamma ray counts (or relative absorption) against the velocity of the source with respect to the absorber, usually measured in millimetres per second, which constitutes the energy axis of this spectroscopy. For the simplest case of both source and absorber containing the Mössbauer isotope in the same cubic environment the spectrum consists of a single absorption line at zero velocity, as shown in Figure 1.3.

The experimental aspects of Mössbauer spectroscopy are well developed and are documented in considerable detail elsewhere (e.g. Gonser, 1975; Greenwood & Gibb, 1971; Gruverman, 1965–74; May, 1971; Wertheim, 1964). A schematic diagram of a typical Mössbauer spectrometer is shown

Fig. 1.2. Nuclear decay scheme of $^{57}$Co showing the transition giving the 14.4 keV Mössbauer gamma ray.

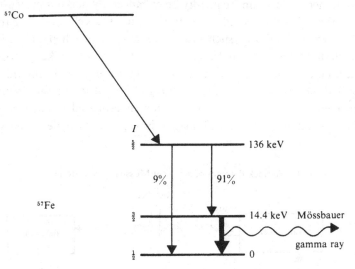

Fig. 1.3. A schematic representation of Mössbauer spectroscopy with the simplest situation of source and absorber nuclei in identical environments and showing the resulting Mössbauer spectrum with an absorption line at zero velocity.

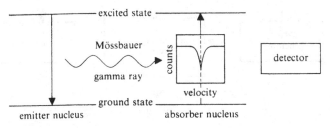

in Figure 1.4. The source motion is normally obtained by an electromechanical transducer, similar to a loudspeaker, and driven by a suitable electronics system. The detector system is dependent on the energy of the particular gamma ray involved but is essentially standard nucleonics with a single channel analyser selecting the counts corresponding to the Mössbauer gamma ray. The synchronisation of the gamma ray counting and the source motion is now generally achieved by a microprocessor system in which the counts are accumulated in channels corresponding to the velocity of the source relative to the absorber. The spectrum is accumulated for a period typically of the order of hours or days during which this spectrum may be monitored on a display screen. When a spectrum with a satisfactory signal-to-noise ratio has been obtained it is normally stored on magnetic tape or disc for subsequent computer analysis.

The quality of the Mössbauer spectrum and also the information that can be obtained from it can frequently be enhanced by making measurements with the absorber (and sometimes the source) under particular experimental conditions, such as low temperatures, applied magnetic fields or high pressures. Many of the experimental aspects of making Mössbauer measurements on various systems are concerned with providing the required sample environment. Although the nature of the absorber is dependent on the type of system being investigated it must contain a sufficient amount of the Mössbauer isotope to provide a measurable

Fig. 1.4. A block diagram of a typical Mössbauer spectrometer.

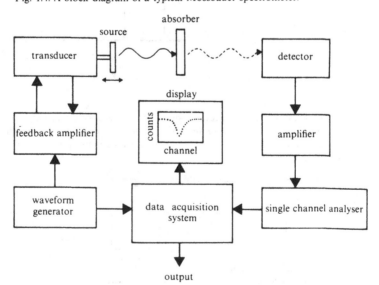

resonant absorption and it must normally be in a solid form (e.g. single crystal, powdered solid, metal or frozen solution).

The experimental implementation of Mössbauer spectroscopy discussed in this section corresponds to the most usual case of an absorption experiment, involving transmission geometry and the detection of the Mössbauer gamma rays. It is also possible to carry out Mössbauer spectroscopic measurements in which the sample under investigation forms the radioactive source, thus allowing measurements involving elements which form the parent nuclei in a Mössbauer decay scheme. Another possibility, which is particularly appropriate for thick samples which are not amenable to examination by transmission, involves a scattering experiment with detection of either the Mössbauer gamma rays, or the conversion electrons or X-rays which are associated with the absorption process. These and other more exotic Mössbauer spectroscopy experiments, which are considered in detail elsewhere (e.g. Gonser, 1981; Gruverman, 1965–74), enable Mössbauer spectroscopy to be applied to a wider range of systems.

## 4 Factors determining the Mössbauer spectrum

A Mössbauer spectrum is characterised by the number, shape, position and relative intensity of the various absorption lines. These features result from the nature of the various hyperfine interactions and their time dependence, as well as on any motion of the Mössbauer nuclei.

The total absorption intensity of the spectrum is a function of the concentration of Mössbauer nuclei in the absorber and the cross-sections of the nuclear processes involved. This absorption intensity, together with the signal-to-noise ratio of the detection system and the total number of counts, determine the quality of the Mössbauer spectrum and the accuracy with which information can be obtained from it. It should be noted that Mössbauer spectroscopy is not generally an appropriate technique for measuring the total concentration of a certain nuclide within a system because it is usually relatively insensitive to small concentrations and because the absorption intensity depends on a number of other factors which may be difficult to quantify. However, relative concentrations of different chemical forms of the Mössbauer nuclide can frequently be obtained, such as the relative concentration of the element in different oxidation states.

The following sections will consider the various hyperfine interactions and other effects which determine the nature of the Mössbauer spectrum, and will also consider briefly how these factors relate to the types of information that can be obtained by Mössbauer spectroscopy.

### *4.1    Isomer shift*

The isomer shift of the absorption lines in the Mössbauer spectrum, also sometimes known as the chemical shift, the chemical isomer shift or the centre shift, is a result of the electric monopole (Coulomb) interaction between the nuclear charge distribution over the finite nuclear volume and the electronic charge density over this volume. This shift arises because of the difference in the nuclear volume of the ground and excited states, and the difference between the electron densities at the Mössbauer nuclei in different materials. In a system where this electric monopole interaction is the only hyperfine interaction affecting the nuclear energy levels, the nuclear ground and excited states are unsplit, but their separation is different in the source and absorber by an amount given by the isomer shift $\delta$.

This situation is shown schematically in Figure 1.5. In this diagram the transition energy between the nuclear ground and excited states in an isolated nucleus (shown on the left of the diagram) is modified by the different electronic environments of the nucleus in the source and absorber. The application of a Doppler velocity to the source is therefore necessary to attain resonance and detect the shift in the position of the resonance line from zero velocity (*cf.* Figure 1.3); the resulting Mössbauer spectrum consists of a single absorption line at a position determined by the isomer shift (Figure 1.5(*a*)). When additional hyperfine interactions are present the isomer shift sets the position of the centre of gravity of the whole Mössbauer spectrum. It should be noted that the isomer shift is not an absolute quantity since it represents the difference between the electric monopole

Fig. 1.5. The effects on the nuclear energy levels of $^{57}$Fe of (*a*) the isomer shift and (*b*) the quadrupole splitting. The Mössbauer absorption transitions and the resulting spectra, which give the isomer shift $\delta$ and the quadrupole splitting $\Delta$, are also shown.

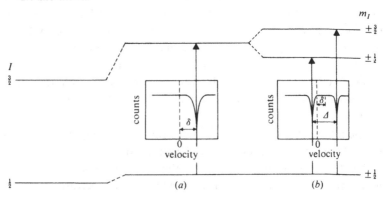

interactions in the source and the absorber. Therefore in order to make comparisons of the isomer shifts obtained from different absorbers the isomer shift data are generally expressed relative to a standard absorber which is also used to determine the zero of the velocity axis of the spectrum. These standard absorbers are specific to the particular isotope used and for the common Mössbauer isotopes there are universally recognised reference standards for isomer shifts. Thermal vibration of the nuclei will also shift the gamma-ray energy as a result of the relativistic second-order Doppler effect. As these thermal vibrations are temperature dependent the temperature of both the source and absorber should be considered when quoting and comparing isomer shifts.

The isomer shift, in common with the other hyperfine parameters of the Mössbauer spectrum, is a function of both nuclear and electronic properties of the system, which are combined in such a way that independent quantitative information on both these properties cannot be obtained by Mössbauer spectroscopy alone. Since it is the electronic properties which are usually of interest and because the nuclear parameters are constant, the hyperfine parameters are most frequently used to compare the electronic properties of different systems. In situations where the nuclear parameters have already been determined it is possible to derive quantitative information on the electronic properties from the Mössbauer spectroscopic measurements. For example, in the case of the isomer shift the relevant nuclear parameters are the nuclear radii of the ground and excited states and when these are known the electronic parameters, which are the electron densities at the Mössbauer nuclei in the source and absorber, can be obtained. The latter are dependent on the electronic structure of the atom and hence the isomer shift $\delta$ is an important means by which atomic oxidation states, which are sometimes difficult to determine by other techniques, can be directly investigated. Similarly covalency effects and the shielding of one set of electrons by another also influences the electronic environment of the nucleus and may be reflected in changes in the isomer shift. The isomer shift data can also be used to make a quantitative assessment of the electron-withdrawing power of substituent electronegative groups and the bonding properties of ligands.

### 4.2   Quadrupole splitting

In considering the electric monopole interaction and the resulting isomer shift it is implicitly assumed that the nuclear charge distribution is spherical. However, nuclei in states with a nuclear angular momentum quantum number $I > \frac{1}{2}$ have non-spherical charge distributions which are characterised by a nuclear quadrupole moment. When the nuclear quadrupole moment experiences an asymmetric electric field, produced by an asymmetric electronic charge distribution or ligand arrangement and characterised by a tensor quantity called the electric field gradient (EFG), an electric quadrupole interaction occurs which gives rise to a splitting of the nuclear energy levels corresponding to different alignments of the quadrupole moment with respect to the principal axis of the electric field gradient.

In the case of $^{57}$Fe the excited state has $I = \frac{3}{2}$, and in the presence of a non-zero electric field gradient this splits into two substates characterised by $m_I = \pm\frac{1}{2}$ and $m_I = \pm\frac{3}{2}$. This situation leads to a two-line spectrum, with the two lines separated by the quadrupole splitting $\Delta$ (Figure 1.5($b$)).

The quadrupole splitting obtained from the Mössbauer measurement involves both a nuclear quantity, the quadrupole moment, and an electronic quantity, the electric field gradient. The value of the nuclear quadrupole moment is fixed for a given nuclide and it is the details of the electric field gradient which can be derived from the Mössbauer spectrum. The electric field gradient contains a number of different contributions. One of these arises from the valence electrons of the Mössbauer atom itself and is associated with asymmetry in the electronic structure. This asymmetry results from partly filled electronic shells occupied by the valence electrons. Another contribution is from the lattice and arises from the asymmetric arrangement of the ligand atoms in non-cubic lattices. Molecular orbitals can also contribute to the electric field gradient. The effects of these contributions at the Mössbauer nucleus are modified by the polarisation of the core electrons of the Mössbauer atom which may reduce or enhance the electric field gradient.

The quadrupole splitting observed in a particular system therefore reflects the symmetry of the bonding environment and the local structure in the vicinity of the Mössbauer atom. Valuable information can be obtained by comparing quadrupole splitting data obtained from related materials, and this can be particularly informative when the data are considered in conjunction with data on the isomer shifts. Quadrupole splitting data can for example give information relating to the electronic population of various orbitals, isomerisation phenomena, ligand structure, short-lived reaction intermediates, semiconductor properties and the defect structure of solids. As with the other Mössbauer hyperfine parameters, the usefulness of quadrupole splitting data for yielding significant information is strongly dependent on whether he nuclear parameters are sufficiently favourable to allow differences in the electronic environment to be reflected in significant and interpretable changes in the spectra.

### 4.3    *Magnetic splitting*

When a nucleus is placed in a magnetic field there is a magnetic dipole interaction between any nuclear magnetic moment and the magnetic field. This interaction completely raises the degeneracy of a nuclear state with an angular momentum quantum number $I > 0$ and splits it into $2I + 1$ substates. In the case of $^{57}$Fe the ground state with $I = \frac{1}{2}$ splits into two substates and the excited state with $I = \frac{3}{2}$ splits into four substates. The $\Delta m_I = 0, \pm 1$ selection rule, appropriate to the Mössbauer gamma ray, leads to six possible transitions and hence a Mössbauer spectrum with six absorption lines. This is shown schematically in Figure 1.6.

Since the splitting of the spectral lines is directly proportional to the magnetic field experienced by the nucleus, Mössbauer spectroscopy provides a very effective means by which this field may be measured. The transition probabilities between the nuclear substates affect the intensities of the lines in the Mössbauer spectrum which can therefore give information on the relative orientation of the magnetic field at the nucleus and the direction of propagation of the gamma-ray beam.

The total magnetic field experienced by the nucleus is a vector sum of the magnetic hyperfine field and any external applied magnetic field. The

Fig. 1.6. The effect of magnetic splitting on the nuclear energy levels of $^{57}$Fe, showing the Mössbauer absorption transitions and the resulting spectrum in the absence of quadrupole splitting. The overall splitting of the lines in the spectrum is proportional to the total magnetic field at the nucleus.

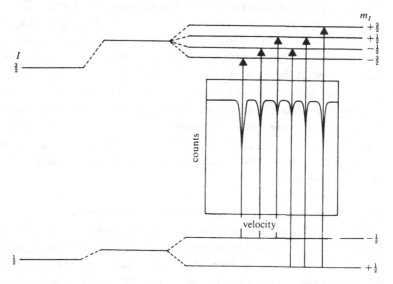

magnetic hyperfine field at the nucleus arises from any unpaired spin of the atoms' own electrons and therefore depends on the oxidation and spin state of the atom. An important feature of the magnetic hyperfine field is that only the unpaired electrons can contribute to it, either directly or indirectly. In principle therefore the situation is much less complex than with isomer shift or quadrupole splitting data, in which contributions may come from all the electrons, both within the atom and also in the molecule and lattice. Magnetic hyperfine fields are observed in the Mössbauer spectra of magnetically ordered systems, or of paramagnetic systems when the electron spin relaxation times are long. Relaxation processes, whereby the electronic spins change direction with a characteristic time, are an important factor in determining the relationship between a magnetic system and its Mössbauer spectra.

The ability to apply external magnetic fields provides an important additional factor in investigating systems with magnetic hyperfine splitting. The total magnetic field at the nucleus then arises from a vector sum of the two contributions and changes in the splitting of the spectral lines following the application of the external field can be of considerable assistance in interpreting the spectrum. The magnetic hyperfine interaction is thus unique among the three Mössbauer hyperfine interactions in being able to be readily modified by an external variable.

The effects of magnetic and quadrupole splitting on the Mössbauer spectrum may be considerably more complex when they are present together, and the observed spectrum is then strongly dependent on their relative magnitudes and orientations. The application of a magnetic field to a system with no unpaired spins, which therefore has no magnetic hyperfine interaction, leads to a magnetic splitting in addition to the quadrupole splitting and this can provide information on the geometry of the electric field gradient at the nucleus.

The data obtained from magnetically split Mössbauer spectra can be used to investigate the magnetic ordering and structure of magnetically ordered systems, the nature of the magnetic interactions, the size of the magnetic moment on particular atoms, and details of the electronic structure of the atom which relate to the magnetic hyperfine field at the Mössbauer nucleus.

### 4.4    *Time-dependent effects, relaxation and dynamics*

The Mössbauer process and the hyperfine interactions have characteristic times and the spectrum observed in any situation depends on whether the properties of the nuclear environment or the position of the nucleus are changing relative to these times. These time-dependent effects can influence both the spectral lineshapes and the values of the Mössbauer hyperfine parameters.

Time-dependent changes in the nuclear environment, often referred to as relaxation processes, can relate to structural changes in systems as a function of time as well as to changes involving the electronic configuration. Such processes can affect all of the hyperfine interactions. Structural changes can be on a macroscopic scale, as in the case of diffusion and melting, or on a localised scale, such as rotation within a molecule. A time dependence of the electronic structure occurs in systems with valence fluctuations and can also result from the after effects of nuclear transformations. When the time dependence involves the orientation of the electronic spin and hence affects the magnetic hyperfine interaction, the process is often known as magnetic relaxation and can be considered in terms of a time dependence of the magnitude and direction of the magnetic hyperfine field experienced by the nucleus.

The influence which relaxation processes and other time-dependent effects have on the Mössbauer spectra is a function of the relative timescales associated with the effects themselves and the timescales of the nuclear transitions and hyperfine interactions. In order to interpret the Mössbauer spectra in terms of time-dependent effects each type of relaxation phenomenon must be considered in the context of the appropriate timescale. Thus, as with the hyperfine interactions themselves, the time dependence results from the interplay of both nuclear and extra-nuclear factors.

Since the Mössbauer effect is intimately related to any motion of the emitting or absorbing nucleus on either a microscopic or macroscopic scale, Mössbauer spectroscopy provides a potential means by which information on nuclear dynamics, and hence on the dynamics of a system in which the Mössbauer nucleus acts as a probe, can be obtained. Any motion of the Mössbauer nucleus can influence the Mössbauer spectrum in two ways. Firstly, because this motion may be related to the vibrational properties of the system it can influence the recoil-free fraction and hence the absorption intensity of the spectrum itself. Since the absolute absorption intensity is dependent on a large number of other factors, which may be difficult to determine accurately, any change in recoil-free fraction is most usefully followed as a function of temperature in order to obtain information on the vibrational properties of the system. The second way in which the effects of any motion of the Mössbauer nucleus in the source or absorber are manifested is in the Mössbauer spectroscopic linewidths, as this motion can be thought of as an additional Doppler motion which may partially smear out the resonant absorption. Since the linewidths are also dependent on other factors it is again often useful to follow them as a function of temperature or other relevant variable in order to separate the various effects.

The dynamic properties of the Mössbauer nucleus which can be monitored as a result of their effect on the Mössbauer spectrum can arise from the lattice dynamics of the solid in which the nucleus is situated. It can also result from the motion of a localised part of the system, such as a molecular motion, or from motion of the whole system within its environment. As the effect on the spectrum depends only on the actual motion of the nucleus and not on its origin it is not possible to distinguish directly between the possible sources of any such motion. Since the motion is often related to the internal energy of the system under investigation it is frequently studied as a function of the various parameters which determine the behaviour of the system, and particularly the temperature.

**5    Applications of Mössbauer spectroscopy**

From its beginnings as an elegant experiment in nuclear and solid state physics Mössbauer spectroscopy has come to be applied to different problems in many areas of science and technology. The interdisciplinary nature of the technique and its numerous applications are illustrative both of the way in which science develops and of how it leads to technological progress. As a means of investigation it successfully complements other experimental methods but has several features which make it an especially powerful technique in a number of important situations and applications.

Many chemical applications of Mössbauer spectroscopy exploit the sensitivity of the technique to changes in electron density at the nucleus. Compounds in which atomic oxidation states and coordination numbers are unclear have been examined by Mössbauer spectroscopy and problems resolved from a consideration of the isomer shift data. Interpretations of this parameter have led to a better understanding of the ionicity and covalency of bonds, the electronic properties of intermetallic, inorganic and organometallic compounds and, by use of molecular orbital calculations, more sophisticated appreciations of the nature of bonding in solids. The chemical uses of quadrupole splitting data for the examination of the site symmetry of an atom and the electronic arrangement around the nucleus have developed in a number of areas, including studies of molecular rotation in solids and phenomena such as spin state crossover, which is observed in certain complexes. Mössbauer spectroscopic studies at high pressures have been used to investigate solid state transformations and changes in chemical bonding and structure which are induced under pressure.

Different phenomena can be investigated when Mössbauer measurements are carried out on magnetic substances. Information on the type of magnetism, the onset of magnetic ordering, the magnetic structure and the magnetisation can be obtained. The Mössbauer spectra of such substances frequently show interesting features in the critical region near magnetic ordering transitions. Recent developments in this area have involved investigations of iron-containing compounds with low-dimensional properties and have shown that Mössbauer measurements of the magnetic hyperfine fields are a powerful means by which the large magnetic fluctuations which occur in these materials can be investigated.

A wide range of solids including glasses, disordered alloys and finely divided samples are amenable to examination by Mössbauer spectroscopy and even dissolved molecules and complexes can be examined when the solutions are frozen. As well as investigating carefully prepared pure solids, the technique has also been widely used in materials science research on technologically important solids which are frequently far from being pure, homogeneous, single-phase materials. It is often the complexity or heterogeneity of these materials which gives rise to their interesting technological properties. Mössbauer spectroscopy can be a good technique for these types of studies because it is specific to the particular nuclide being investigated, which thus provides a simplifying factor.

Although the diversity of potential applications of Mössbauer spectroscopy is too extensive to detail here, it is relevant to mention a few of the fields in which the technique has been found to be particularly useful. Metallurgy has proved to be an important area which has benefited from the application of this technique, indeed [57]Fe Mössbauer spectroscopy has developed as a valuable experimental method in steel research and commercial operation as a means by which the iron phase distributions both in ores and in processed steel can be monitored.

Catalysis is a multi-disciplinary area of science and within this large field several regions of interest have been found to be amenable to investigation by Mössbauer spectroscopy. A particularly successful application of Mössbauer spectroscopy within this field has involved the *in situ* study of solids which catalyse gaseous reactions. The development of special cells in which these catalysts may be studied in their authentic working environment is an important feature of this work. In many of the Mössbauer spectroscopic studies of catalysts the data have been used in conjunction with that from other techniques to elucidate fundamental properties of both the bulk and surface which influence catalytic performance.

The study of solid surfaces has provided another area of application in which Mössbauer spectroscopy has made a significant contribution. The power of the technique in this area has been enhanced by the development of conversion electron Mössbauer spectroscopy in which the decay of excited nuclei in the absorber is monitored by detecting the internally converted electrons. Because of the shallow escape depth of these electrons only the surface layers are studied. Investigations using this technique have involved the oxidation and corrosion of iron alloys in a variety of gaseous atmospheres and also studies of the various chemical surface coatings of metals which are used to produce corrosion resistance and other modifications of the surface. Thin films, small particles, ion implanted alloys, minerals and archaeological materials have also been investigated using conversion electron Mössbauer spectroscopy.

The use of Mössbauer spectroscopy to study biological systems is an important area of application which derives mainly from the sensitivity of iron to Mössbauer spectroscopic investigation and the fundamental role of this element in many biomolecules. The specificity of the technique enables detailed examination of the local environment of iron in large and complex biological materials such as proteins and enzymes in which the iron atoms are at the active centres where the biologically significant changes occur. Such studies have led to an enhanced understanding of the specific functions of certain complex molecules in biological systems. There have also been medical applications of Mössbauer spectroscopy, involving studies of tissue and blood samples, which have led to the identification of important pathological factors.

Mössbauer spectroscopy has also found considerable applications in the investigation of mineral systems. Much of this work has involved the determination of the number and structure of the chemical phases within certain minerals and has been of both scientific and technological importance. The importance of small amounts of doping and impurities in affecting the Mössbauer spectrum provides a significant feature of much of this work.

# 2

# Mössbauer spectroscopy and the chemical bond

R. V. PARISH

## 1 Introduction

The creation of a chemical bond involves the sharing of two or more electrons between a pair of atoms. If the atoms are identical, as in a homonuclear diatomic molecule, there is little change in the total electron density on each atom, but there may be a slight redistribution of electrons between the valence orbitals to achieve a more satisfactory hybridisation. If the atoms differ, the bond will be more or less polar, depending on the disparity in electronegativities: one atom will gain electron density at the expense of the other. This effect is at its most extreme in the formation of an ionic bond. When an atom forms more than one bond, considerable rehybridisation may take place, with or without gain or loss of electron density by the electronegativity effect. An examination of the distribution of electron density about a particular atom can therefore give considerable information about the bonds involving that atom.

Mössbauer spectroscopy is well suited to such a study, since two major parameters of a spectrum can be related directly to the populations and changes in population of the valence shell orbitals. The isomer shift relates to the total electron density on the atom, and the quadrupole splitting reflects any asymmetry in the distribution of electron density. Thus, each parameter is capable of giving information about bonding: from their combination, it is often possible to make quite detailed analyses.

## 2 Isomer shift as a probe for bonding
### 2.1 Isomer shift and electron density

The isomer shift $\delta$ is directly proportional to the difference in total electron density at the nucleus between the absorber and the standard,

$$\delta = \alpha[|\psi_A(0)|^2 - |\psi_S(0)|^2] \tag{2.1}$$

where the constant of proportionality $\alpha$ depends principally on the relative change in the nuclear radius between the excited state and the ground state $\delta R/R$. For most nuclei the excited state has the larger radius, but for some, notably $^{57}$Fe, the reverse is true. Differences in isomer shift between compounds therefore reflect, with varying sensitivity, differences in electron density at the nucleus, which in turn should indicate differences in the populations of the valence shell orbitals.

The major contribution to the total electron density at the nucleus comes from the $1s$ electrons. Other $s$ electrons make lesser contributions as the principal quantum number increases, while $p$ and $d$ electrons exert shielding effects. Table 2.1 shows values calculated for iron atoms and ions with various electron configurations. The total contribution due to the core electrons is quite insensitive to changes in the population of the valence shell orbitals. Obviously the electron density due to the $4s$ electrons depends directly on the population of the $4s$ orbital, but it is also significantly affected, in the opposite sense, by the $3d$ population (Figure 2.1(a)). This is a perfectly general situation; increase in the valence shell $s$ population increases the electron density at the nucleus, while increase in the $p$ or $d$ populations causes a decrease (Figure 2.1(b)). Because the latter

Table 2.1. *Calculated electron densities at the $^{57}$Fe nucleus*[a]

| | Fe⁰, [Ar]$3d^7 4s^1$ | | Fe⁺, [Ar]$3d^6 4s^1$ | | Fe⁺, [Ar]$3d^7$ | |
| --- | --- | --- | --- | --- | --- | --- |
| | | Cumulative total | | Cumulative total | | Cumulative total |
| $1s^2$ | 170 877.96 | | 170 874.52 | | 170 878.87 | |
| $2s^2$ | 16 193.00 | 187 070.96 | 16 191.30 | 187 065.82 | 16 192.01 | 187 070.88 |
| $2p_{\frac{1}{2}}^2$ | 87.01 | 187 157.97 | 86.98 | 187 152.80 | 87.01 | 187 157.89 |
| $3s^2$ | 2205.47 | 189 363.44 | 2229.97 | 189 382.77 | 2198.54 | 189 356.44 |
| $3p_{\frac{3}{2}}^2$ | 11.28 | 189 374.22 | 11.47 | 189 394.24 | 11.26 | 189 367.70 |
| $4s^1$ | 38.86 | | 70.06 | | | |

[a] The data in the table include contributions from $p$ electrons because it is necessary to use the relativistic quantum mechanical calculations developed by Dirac (1928), in which quantum numbers are differently defined, and electron distributions are slightly different: in particular, $p$ orbitals do not have nodes at the nucleus. The relativistic treatment also gives higher electron densities at the nucleus than the conventional treatment, reflecting the increase in speed and, hence, mass of the electron as it approaches the nucleus; the electron therefore spends more time close to the nucleus than expected. Relativity effects become more important as the nuclear charge increases (Pitzer, 1979; Pyyko & Desclaux, 1979; Powell, 1968; these references also give interesting discussions of the chemical consequences of relativity effects).

(From Trautwein, Freeman & Desclaux, 1976.)

Fig. 2.1. Changes in electron density at the nucleus for free ions as a function of electron configuration for (a) iron (core level 189 367.70 $a_0^{-3}$), and (b) tin (core level 2 334 530.0 $a_0^{-3}$). (Data from Trautwein, Freeman, & Desclaux, 1976, and from Ellis, D. E., quoted by Ruby & Shenoy, 1978.)

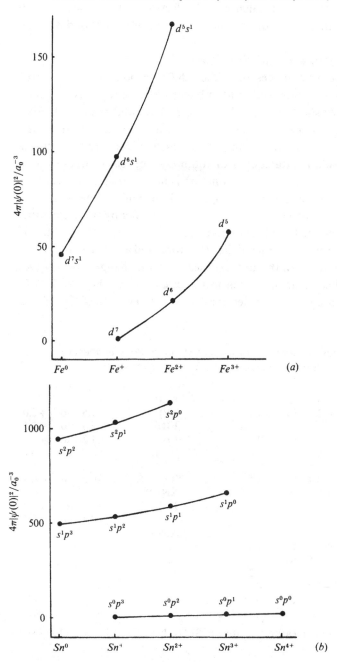

effect operates indirectly, by shielding the $s$ electrons, changes in the $p$ or $d$ populations have much less pronounced effects on the isomer shift.

The isomer shift thus responds to any factor which can change the number or distribution of valence shell electrons. It therefore reflects changes such as those in oxidation state, hybridisation and bond polarity.

**2.2    *Isomer shift and oxidation state***

Characteristic ranges of isomer shift are often associated with particular oxidation states of the Mössbauer atom, particularly for the $p$ block elements. As shown in Table 2.2, there is a large and distinct difference in the ranges for the penultimate and the highest oxidation state. These states differ by two units, and correspond to the removal of two $s$ electrons. Thus, the two oxidation states of tin or antimony are easily distinguished by isomer shift data alone, and the distinction is not blurred by the range of values produced by variation in the ligands. Tellurium, iodine, and xenon also have a series of lower oxidation states, which differ by two $p$ electrons, down to the closed shell configurations of $Te^{2-}$, $I^-$ and $Xe^0$. The addition of $p$ electrons results in a shielding of the $s$ electrons, and a return of the isomer shift towards the value for the fully oxidised state; the changes are, however, much less marked than those due to direct change in the number of $s$ electrons. Very roughly, one $p$ electron has about one-sixth of the effect of

Table 2.2. *Ranges of isomer shift (in mm/s) for p block elements in various oxidation states (organic derivatives are excluded)*[a]

| | | | | | | | |
|---|---|---|---|---|---|---|---|
| $^{119}$Sn(IV) | −0.5 | to | +1.5 | $^{129}$I(VII) | −2.3 | to | −4.5 |
| Sn(II) | +2.2 | to | +4.2 | I(V) | +1.5 | to | +3.0 |
| | | | | I(III) | +2.8 | to | +3.5 |
| $^{121}$Sb(V) | −2 | to | +12 | I(I) | +1.2 | to | +1.9 |
| Sb(III) | −8 | to | −2 | I(−I) | −0.5 | to | −0.2 |
| | | | | | | | |
| $^{125}$Te(VI) | −0.9 | to | −1.4 | $^{129}$Xe(VIII) | −0.1 | to | −0.2 |
| Te(IV) | +0.7 | to | +2.0 | Xe(VI) | 0.0 | to | +0.7 |
| Te(II) | +0.3 | to | +1.3 | Xe(IV) | +0.2 | to | +0.9 |
| Te(−II) | | ca. | −0.1 | Xe(II) | −0.1 | to | +0.5 |
| | | | | Xe(0) | | 0 | |
| $^{127}$I(VII) | +0.8 | to | +1.5 | | | | |
| I(V) | −0.5 | to | −1.0 | | | | |
| I(III) | −1.0 | to | −1.2 | | | | |
| I(I) | −0.4 | to | −0.7 | | | | |
| I(−I) | +0.1 | to | +0.05 | | | | |

[a] The isomer shifts are measured relative to $SnO_2$, InSb, I/Cu, ZnTe and Xe/BHC respectively.

one *s* electron; the closed-shell isomer shift values are thus half way between those for the formal $ns^2$ and $ns^0$ configurations.

The various isotopes all show different scales of sensitivity of isomer shift to electron configuration, since all have different values of $\delta R/R$; for $^{121}$Sb and $^{127}$I, $\delta R/R$ is negative and the scales are reversed. The full expression for the isomer shift is given by

$$\delta/\mathrm{mm\ s^{-1}} = 12.158 Z (E_0/\mathrm{keV})^{-1}(\delta R/R)(R^2/\mathrm{fm}^2)$$
$$\times \left[ |\psi_A(0)|^2 - |\psi_S(0)|^2 \right]/a_0^{-3} \qquad (2.2)$$

where $Z$ is the atomic number, $E_0$ is the gamma-ray energy, $R$ is the nuclear radius and $a_0$ is the first Bohr radius. The factors appearing in front of the square brackets in Equation (2.2) correspond to the proportionality constant $\alpha$ in Equation (2.1). The values of these factors for various Mössbauer isotopes are listed in Table 2.3.

For the elements from tin to xenon the most important single factor affecting the isomer shift is $\delta R/R$, and this accounts for the high sensitivity for $^{121}$Sb and $^{129}$I. Since $\delta R/R$ cannot be measured directly, it is not possible to quantify the relationship between isomer shift and electron configuration without making assumptions about the electronic populations in particular systems. The difficulties of such calibration procedures are discussed later.

For the transition metals changes in oxidation state involve alteration in the number of *d* electrons which, other things being equal, would severely

Table 2.3. *Nuclear factors affecting the isomer shift*[a]

|  | $Z$ | $E_0/\mathrm{keV}$ | $(\delta R/R)2R^2/\mathrm{fm}^2$ | $\alpha$ | Relative $\alpha$ | $\delta\|\psi(0)\|^2$ |
|---|---|---|---|---|---|---|
| $^{119}$Sn | 50 | 23.9 | +3.3 | +41.9 | 1.0 | 4.1[b] |
| $^{121}$Sb | 51 | 37.1 | −24.6 | −206 | −4.9 | 5.3[b] |
| $^{125}$Te | 52 | 35.5 | +2.7 | +24.0 | +0.57 | 6.4[b] |
| $^{127}$I | 53 | 57.6 | −14.4 | −80.6 | −1.9 | 7.5[b] |
| $^{129}$I | 53 | 27.7 | +18.5 | +215 | +5.1 | 7.5[b] |
| $^{129}$Xe | 54 | 39.6 | +1.5 | +12.5 | +0.30 | 8.8[b] |
| $^{57}$Fe | 26 | 14.4 | −14.3 | −157 | −1.0 | 37[c] |
| $^{99}$Ru | 44 | 89.4 | +20.0 | +60.0 | +0.38 | 64[c] |
| $^{193}$Ir | 77 | 73.0 | +5.5 | +35.2 | +0.22 | 440[c] |
| $^{197}$Au | 79 | 77.3 | +8.6 | +53.3 | +0.34 |  |

[a] The symbols are the same as those used in Equations (2.1) and (2.2).
[b] Difference between $|\psi(0)|^2$ for the configurations $5s^1 5p_{\frac{3}{2}}^2$ and $5s^1 5p_{\frac{3}{2}}^1$ (Ellis, D. E., quoted by Ruby & Shenoy, 1978).
[c] Difference between $|\psi(0)|^2$ for the configurations $nd^6$ and $nd^5$ (Freeman & Ellis, 1978).

reduce the sensitivity of the isomer shift. Fortunately, the most widely studied isotope, [57]Fe, has a relatively large value for $\alpha$ (Table 2.3), so that the oxidation state ranges are easily observed (Table 2.4), and diagnosis is again simple. Since the isomer shift becomes more positive as the number of $d$ electrons increases, i.e. as the $s$ electron density becomes more shielded, it is evident that $\delta R/R$ is negative.

The data to the left-hand side of Table 2.4 refer to ionic compounds in which the iron ions have a high spin (spin free) configuration. With ligands sufficiently high in the spectrochemical series the crystal field is large, giving spin pairing and a low spin configuration. This is reflected in the isomer shift values, which now cover much narrower ranges, and the sharp distinction between oxidation states is lost (Table 2.4). In general, spin pairing in iron compounds produces a decrease in the isomer shift, i.e. an increase in $s$ electron density at the nucleus. This is due to the increase in covalency of the metal–ligand bonds, with a corresponding increase in the $4s$ electron density. The accumulation of negative charge on the metal atom also produces an increase in the effective radius of the $d$ orbitals (the nephelauxetic effect), and a deshielding. There may also be delocalisation of $d$ electrons through $\sigma$-antibonding, $\pi$-bonding or $\pi$-antibonding molecular orbitals. Similar considerations apply to [61]Ni, except that $\delta R/R$ is very small. Under these conditions, the second-order Doppler shift is comparable with the isomer shift and both are much smaller than the linewidth.

All other transition metals for which the Mössbauer effect can be observed belong to the second and third series, and show only spin paired configurations in all oxidation states. In addition, $\alpha$ is small for many of the isotopes and the isomer shift ranges are considerably smaller than the linewidths. The most informative nuclei are [99]Ru, [193]Ir, and [197]Au, and the isomer shift ranges for these are shown in Table 2.5. For the higher

Table 2.4. *Isomer shift ranges (in mm/s) in iron compounds*[a]

|  | High spin | | | Low spin | | |
|---|---|---|---|---|---|---|
| Fe(VI) | $-0.8$ | to | $-0.9$ | | | |
| Fe(IV) | $-0.2$ | to | $+0.2$ | $+0.1$ | to | $+0.2$ |
| Fe(III) | $+0.1$ | to | $+0.5$ | $-0.1$ | to | $+0.5$ |
| Fe(II) | $+0.6$ | to | $+1.7$ | $-0.2$ | to | $+0.5$ |
| Fe(0) | | | | $-0.2$ | to | $-0.1$ |
| Fe($-$II) | | | | | *ca.* | $-0.2$ |

[a] Relative to iron metal.

oxidation states of ruthenium and iridium, very few data are available, but for the lower oxidation states, complexes involving wide varieties of ligands have been studied, and the isomer shift ranges are correspondingly broad. The trends are best seen by comparing sets of compounds containing similar ligands (Table 2.6).

## 2.3    *Isomer shift and the nature of the ligand*
### 2.3.1    *Main group isotopes*

In molecular systems or coordination complexes, in which the bonding is essentially localised and when there are no complications from non-bonding electrons, the isomer shift reflects the polarity of the bonds. To

Table 2.5. *Isomer shift ranges (in mm/s) for heavy transition metals*[a]

| | | | | | | |
|---|---|---|---|---|---|---|
| $^{99}$Ru(VIII) | | *ca.* 1.1 | $^{193}$Ir(VI) | | *ca.* | 1.5 |
| Ru(VII) | | *ca.* 0.8 | Ir(V) | 0.0 | to | 0.3 |
| Ru(VI) | | *ca.* 0.3 | Ir(IV) | $-0.9$ | to | $-1.3$ |
| Ru(IV) | $-0.2$ to | $-0.3$ | Ir(III) | $+0.4$ | to | $-2.2$ |
| Ru(III) | $-0.4$ to | $-0.8$ | Ir(I) | $+0.3$ | to | $-0.3$ |
| Ru(II) | $-0.2$ to | $-0.8$ | | | | |
| | | | $^{197}$Au(III) | 0.0 | to | $+6$ |
| | | | Au(I) | $-0.5$ | to | $+6$ |

[a] Relative to the elemental metals.

Table 2.6. *Isomer shifts for various transition metals as a function of oxidation state and ligand*[a]

| Complex | $\delta$/mm s$^{-1}$ | Complex | $\delta$/mm s$^{-1}$ |
|---|---|---|---|
| [Fe(phen)$_3$](ClO$_4$)$_3$ | 0.05[b] | K$_3$[Fe(CN)$_6$] | $-0.124$[c] |
| [Fe(phen)$_3$](ClO$_4$)$_2$ | 0.31 | K$_4$[Fe(CN)$_6$] | $-0.047$ |
| K$_2$[RuCl$_6$] | $-0.37$ | | |
| trienH$_3$[RuCl$_6$] | $-0.70$ | [Ru(NH$_3$)$_6$]Cl$_3$ | $-0.49$ |
| | | [Ru(NH$_3$)$_6$]Cl$_2$ | $-0.92$ |
| IrF$_6$ | 1.45 | | |
| IrF$_5$ | 0.06 | | |
| K$_2$[IrF$_6$] | $-1.41$ | K$_2$[IrCl$_6$] | $-0.95$ |
| | | K$_3$[IrCl$_6$] | $-2.26$ |
| K[AuCl$_4$] | 2.05 | K[Au(CN)$_4$] | 5.4 |
| Ph$_4$As[AuCl$_2$] | 1.72 | K[Au(CN)$_2$] | 4.36 |

[a] Except as noted otherwise, data are from the compilations of Wagner & Wagner (1978) and Parish (1982*a, b*, 1984*b*).
[b] From Berrett, Fitzsimmons & Owusu (1968).
[c] From Kerler & Neuwirth (1962).

illustrate this some further restriction on the classes of compounds compared is necessary, since changes in coordination number or stereochemistry can affect the $s$ character of the bonds and hence the isomer shift.

For tin(IV) halides $SnX_4$ ($X = Cl$, Br, I; $SnF_4$ is not isostructural) and the hexahalostannates(IV) $SnX_4Y_2^{2-}$ (X, Y = F, Cl, Br, I) there are good linear relationships between the isomer shift and the electronegativity of the ligands (Figure 2.2). The correlation is negative, showing that the bonding electron density is progressively concentrated on the tin atom as the electronegativity of the ligands decreases. Relationships of this type have been used to derive effective electronegativity values for other groups (Parish & Rowbotham, 1971), e.g. the isomer shift values for $Ph_4Sn$ (1.15 mm/s) and $Bu_4Sn$ (1.35 mm/s) correspond to effective Pauling electronegativities of 2.81 and 2.75 for the phenyl and *n*-butyl groups respectively.

Similarly good correlations are found for the series $Bu^nSnX_nY_{5-n}$ (X, Y = F, Cl, Br; Figure 2.2). However, with the above electronegativity value for the butyl group the data do not coincide with those for the other octahedral

Fig. 2.2. Isomer shift (relative to $SnO_2$) against average Pauling electronegativity $\chi_P$ of the ligands for tin(IV) compounds. A value of 2.75 has been taken for the electronegativity of the *n*-butyl group (see text).

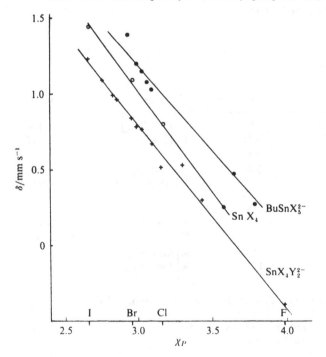

species. To bring the two sets of data together would require an effective electronegativity of about 1.0 for the butyl group, which is far lower than any other treatment would suggest. Clearly, the conventional scales of electronegativity or hardness (Parr & Pearson, 1983) are misleading in this context, since they place alkyl and phenyl groups between iodine and bromine, as indeed do the isomer shift data for $R_4Sn$ and $SnX_4$. The Mössbauer data indicate that in $RSnX_5{}^{2-}$ the organic group places much more electron density on the tin atom than the conventional scales would suggest. However, the isomer shift relates principally to the (tin) $5s$ electron density, and is much less responsive to $5p$ electron density. The anomalously high isomer shift must therefore be interpreted in terms of a fundamental change in hybridisation of the tin atom, which concentrates $5s$ character in the bond to carbon. Given the apparently low effective electronegativity of the organic group, this is in line with Bent's rule (Bent, 1961), which suggests that $s$ character is concentrated in the bond to the least electronegative ligand. This result also accords with chemical experience, which shows that the Sn—C bond is considerably less polar than the Sn—I bond.

A wide range of compounds involving one organic group show good correlation of the isomer shift with the electronegativity of the ligands (Barbieri, Pellerito, Bertazzi & Stocco, 1973). When the number of organic groups is greater, the sensitivity of the isomer shift to ligand electronegativity becomes much less, and may even be lost (Figure 2.3).

Fig. 2.3. Isomer shifts (relative to $SnO_2$) of di- and tri-organotin(IV) compounds as a function of the electronegativity of the halogen.

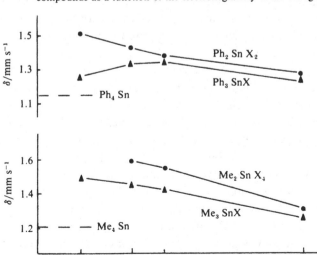

Comparison is more difficult in these series, because many of the compounds are polymeric in the solid state, and the extent of association may differ even within one series. However, it is clear that the organic groups are exerting a buffering effect on the isomer shift out of all proportion to their relative numbers. This effect is again due to the concentration of $s$ character in the Sn—C bonds, so that the change in polarity of the Sn—X bonds affects mainly the $p$ orbital population. A further consequence of this rehybridisation is that the isomer shift is quite sensitive to change in the organic groups, and usually increases with increasing inductive effect: $Ph < Me < Et < Pr^n < Bu^n < Bu^i$.

Differences between various series of organotin compounds are also noticeable. As Figure 2.3 shows, for a given halogen the isomer shift increases in the order $R_4Sn < R_3SnX < R_2SnX_2$. In large measure, this trend is due to changes in the structure, and hence in the hybridisation. For instance, for $X = F$ the Sn—C bond angle increases from $109°$ to $180°$; a parallel increase in the $s$ character of the Sn—C bond is expected, and accounts for the rise in the isomer shift. Similarly, for six-coordinate compounds of the type $R_2SnL_4$ ($L =$ electronegative ligand), there is a distinct difference in the isomer shift between *cis* and *trans* structures, which give ranges of 0.7–1.0 and 1.2–1.6 mm/s respectively (Parish, 1972). This argument is supported by the nuclear magnetic resonance (NMR) coupling constants which in part reflect the $s$ character of the Sn—C bond (Table 2.7).

The effect of changes in coordination number can to some extent be isolated from the other factors affecting the isomer shift. As Figure 2.2 shows, the isomer shift of $SnX_6^{2-}$ is about 0.2 mm/s less than that of $SnX_4$. It might be expected that the coordination of extra ligands would increase the total charge density on the tin atom, and lead to an increase in the isomer shift. However, an increase in coordination number leads to a lengthening of the metal–ligand bonds, decreasing the extent of donation to tin. At the same time, there is an increase in $p$ character (and possibly $d$ character) of the hybrid orbitals on tin. In organotin systems, this results in increased shielding of the $s$ electron density (concentrated in the Sn—C bonds), which explains why the *trans* $R_2SnX_4$ complexes have lower isomer shift values than $R_2SnX_2$. For a given coordination number, the isomer shift appears to be an additive function of the ligands, provided that all ligands are similar in electronegativity or hardness, i.e. all bonds are similar in degree of covalency. In mixed ligand systems, the inequivalence of the hybrid orbitals to different types of ligand destroys the additivity.

The isoelectronic antimony(V) compounds show rather similar trends, when the magnitude and negative sign of $\delta R/R$ are taken into account. As shown in Figure 2.4, the isomer shift becomes more negative in the series

Table 2.7. *Mössbauer and NMR data for organotin(IV) compounds*

| | $\delta$/mm s$^{-1}$ | $\Delta$/mm s$^{-1}$ | C—Sn—C/° | $^2$J($^{119}$Sn—CH)/Hz | $^1$J($^{119}$Sn—$^{13}$C)/Hz |
|---|---|---|---|---|---|
| Me$_4$Sn | 1.20[a,b] | 0.00[a,b] | 109 | 54.0[b,c] | −340[d,e] |
| Me$_3$SnCl | 1.47[a,b] | 3.32[a,b] | 117[f] | 59.7[b,c] | −380[d,g] |
| Me$_3$SnI | 1.43[b,h] | 2.91[b,h] | | | |
| | 1.42[h,i] | 2.69[h,i] | | 58.5[h,j] | |
| Me$_2$SnCl$_2$ | 1.56[a,b] | 3.56[a,b] | 123[k] | 71.0[b,c] | −566[d,l] |
| MeSnI$_3$ | 1.58[b,h] | 1.68[b,h] | | | |
| | 1.56[h,m] | 1.61[h,m] | | 73.0[h,j] | |
| Me$_2$Sn(oxin)$_2$ | 0.88[b,n] | 2.02[b,n] | 111[o] | 71.2[p,q] | |
| Me$_2$Sn(acac)$_2$ | 1.16[b,r] | 4.02[b,r] | 180[s] | 99.0[t] | −966[q,t] |

[a] Averaged from the compilation by Smith (1970).
[b] For the pure compound.
[c] From Holmes & Kaesz (1961).
[d] From McFarlane (1967).
[e] 90% in dioxan.
[f] From Hossain et al. (1979).
[g] 90% in benzene.
[h] From Marks, Drago, Herber & Potasek (1976).
[i] 0.86 mol dm$^{-3}$ in butyl benzene.
[j] 0.03 mol dm$^{-3}$ in methyl cyclohexane.
[k] From Davies, Milledge, Puxley & Smith (1970).
[l] 50% in acetone.
[m] 0.46 mol dm$^{-3}$ in butyl benzene.
[n] From Parish & Platt (1970).
[o] From Schlemper (1967).
[p] From McGrady & Tobias (1965).
[q] In CDCl$_3$.
[r] From Bancroft & Sham (1974).
[s] From Miller & Schlemper (1973).
[t] From Mitchell (1973).

$SbX_5$, $PhSbX_5^-$, $Ph_2SbX_4^-$, and there is a good correlation with the electronegativity of X. The three sets of data show very similar slopes, and there is no evidence of a buffering effect as the number of phenyl groups increases. Presumably, this indicates that the more electronegative antimony atom makes less use of the $5s$ orbital than tin; as will be seen below, this trend is continued by tellurium and iodine.

For tin(II) and antimony(III), the situation is complicated by the presence of the lone pair of electrons, and by the wide variety of structures adopted. This is illustrated in Figure 2.5 for the tin(II) halides and chalcogenides. The two sets of compounds involve different coordination numbers for tin of three and four respectively. There are overall negative correlations of the isomer shift with electronegativity, but the trends are irregular, and confused by the occurrence of two crystal modifications for $SnF_2$ and SnO, which each have different values of the isomer shift.

When the structures are more regular, as in the antimony(III) halides and chalcogenides, there is a better correlation (positive because $\delta R/R$ is negative for $^{121}Sb$), as there is also for the anions $SnX_2Y^-$ (X, Y = halogen).

Fig. 2.4. Isomer shifts of antimony(V) compounds relative to InSb. The right-hand axis refers to the upper plot only.

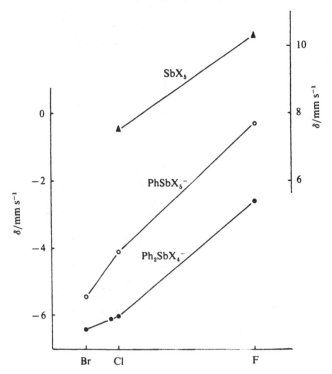

These data are shown in Figure 2.6 and in all these cases the metal has trigonal–pyramidal stereochemistry. The fluorides are again anomalous.

The more electronegative elements, tellurium and iodine, show different behaviour. For the tellurium(IV) and iodine(I) halides, the correlations between isomer shift and electronegativity are positive (Figure 2.7). Since $\delta R/R$ is positive in both cases, this indicates that the $s$ electron density at the nucleus increases with increasing electronegativity of the halogen, which can only be explained on the assumption that the $5s$ orbital is very little, if at all, involved in the bonding, and the changes in isomer shift reflect the decreasing population of the $5p$ orbital. Similar considerations apply to $I^+$, except that there are now three non-bonding pairs of electrons. Greatly reduced sensitivity of the isomer shift to electronegativity is again found in the organic derivatives, e.g. $R_2TeX_2$ (Figure 2.7($b$)), but the substantial quadrupole splitting suggests that the lone pair may have some $p$ character in this case.

The decreasing participation of the $5s$ orbital in the bonding in the series from tin to iodine is presumably caused by the increase in nuclear charge and concomitant increase in binding energy, but the $5p$ electrons are less affected (Table 2.8).

Fig. 2.5. Isomer shifts (relative to $SnO_2$) for tin(II) halides and chalcogenides.

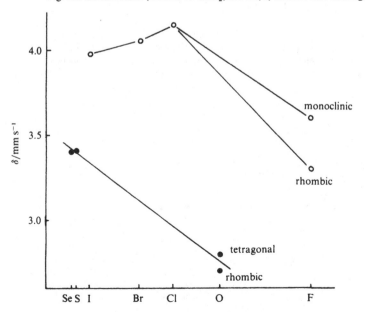

Fig. 2.6. Isomer shifts for systems with pyramidal geometry: (*a*) antimony(III) halides and chalcogenides, and (*b*) tin(II) complexes, $Et_4N(SnX_2Y)$ (X, Y = F, Cl, Br, I). The isomer shifts are relative to InSb for the antimony compounds and relative to $SnO_2$ for the tin complexes.

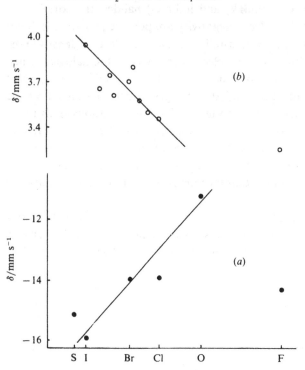

Fig. 2.7. Isomer shifts for iodine(I) and tellurium(IV) compounds: I–X, $TeX_6^{2-}$, $ArTeX_3$, and $Ar_2TeX_2$ (X = F, Cl, Br, I; Ar = aryl). The isomer shifts are relative to ZnTe for the iodine compounds and to I/Cu for the tellurium compounds.

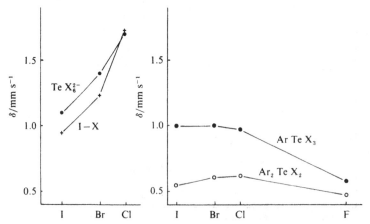

## 2.3.2 Transition elements

In compounds and complexes of the transition elements, non-bonding valence electrons are usually present, but in general they are neither stereochemically active nor involved in the hybridisation schemes. The isomer shift ranges then usually reflect the total number of $d$ electrons (i.e. the oxidation state) and the spin state, as discussed earlier. The values within these broad ranges are governed by the nature of the ligands, i.e. by the charge formally donated by the ligands to the metal ion.

For high spin iron(II) and iron(III) there is a good correlation between the isomer shift and the electronegativity of the ligand (Figure 2.8). These halides are usually regarded as ionic lattices, but they can also be treated as coordination complexes in which each metal ion has $d^2sp^3$ hybridisation and is surrounded by six halide ligands. (The variation in isomer shift shows that these materials are not purely ionic.) The amount of charge donated by the halides presumably increases as their electronegativity decreases. The positive slopes of the correlations imply that electron density donated to the

Table 2.8. *Calculated binding energies (in MJ/mol)*

|        | Sn    | Sb    | Te    | I     |
|--------|-------|-------|-------|-------|
| 5s     | 2.525 | 3.083 | 3.713 | 4.352 |
| 5p     | 1.405 | 1.774 | 1.902 | 2.137 |

From Fischer (1977).

Fig. 2.8. Isomer shifts (relative to iron metal) for halides of (a) iron(II) and (b) iron(III).

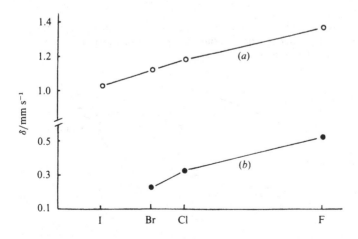

4s orbital has a greater effect on the isomer shift than that donated to the 3d or 4p orbitals.

The slight irregularities in these plots probably reflect differences in crystal structure: for instance, $FeCl_2$, $FeBr_2$ and $FeI_2$ have regular six-coordination ($CdCl_2$ or $CdI_2$ structures), but $FeF_2$ has the less regular rutile structure. The influence of structure is shown in Figure 2.9, where the isomer shift of complex iron(II) fluorides and iron(III) oxides are seen to be positively correlated with the iron–ligand bond distance. Thus, as the bonds become longer, less negative charge is placed on the iron atoms. This effect makes a significant contribution to the dependence of the isomer shift on the coordination number, since the bonds lengthen as the coordination number increases.

The linear correlations of Figure 2.8 suggest that in isostructural series the isomer shift is an additive function of the charge donated by the individual ligands. As will be shown later, this additivity extends to low spin systems also, and to the other transition metals. Such an observation implies, in turn, that little rehybridisation occurs, in marked contrast to the organotin and organoantimony derivatives discussed above.

All transition metals appear to show the same trends with change of ligand. For instance, a change from six halide ligands to six N-donors to six cyanide ligands gives a consistent increase in the electron density at the nucleus (Table 2.9), reflecting the conventional view of the donor abilities of these ligands. Unfortunately, too few data are available to carry this analysis further by direct comparison of related complexes. However, the relative donor powers of ligands can be assessed by the partial isomer shift method discussed in the next section.

The sensitivity of the isomer shift to the chemical environment increases dramatically from iron(II) to ruthenium(II) to iridium(III). As Table 2.2

Fig. 2.9. Isomer shift against iron–ligand distance $d$ for ($a$) iron(II) fluoride systems, and ($b$) iron(III) oxide systems. (Reproduced with permission from Ingalls, van der Woude, & Sawatzky, 1978.)

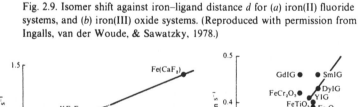

demonstrates, this trend cannot be attributed to the nuclear parameters making up the factor. Rather, it is a function of the increasing total electron density at the nucleus as the principal quantum number of the valence shell increases, and of the greater absolute change in electron density with valence shell population. It is likely that there is also a contribution from the greater covalency of the metal–ligand bond with the heavier metals.

## 2.4    Partial isomer shifts

As indicated above, in systems where the bonding is not highly covalent, the isomer shift appears to be an additive function of the ligands. That is, a partial isomer shift value (p.i.s) can be assigned to each ligand, such that the value of the isomer shift for the complex is given by the sum over all the ligands. The success of such a treatment requires that the isomer shift is relatively insensitive to small differences in bond angle and distance, and that the bonding of any given ligand is insensitive to the nature of the other ligands (Bancroft, Mays & Prater, 1970). This is tantamount to assuming the absence of significant rehybridisation as the ligands are changed, which in turn implies a relatively narrow range of covalencies. Scales of p.i.s. can be set up for many transition metals, to which these criteria apply well, but not for main group elements, to which they do not.

The first scale to be derived was for low spin iron(II) (Table 2.10) for which data for a wide variety of complexes are available (Bancroft *et al.*, 1970; Bancroft, Garrod & Maddock, 1971; Bancroft & Libbey, 1973).

Table 2.9. *Comparison of parameters for* $^{57}$Fe, $^{99}$Ru *and* $^{191}$Ir

|  | $\delta$/mm s$^{-1}$ | Difference in $\delta$/mm s$^{-1}$ | $\alpha^a$ | $\delta|\psi(0)|^{2}$ [b] |
|---|---|---|---|---|
| $[Fe(phen)_3]^{2+}$ | $+0.31^c$ | $-0.36$ | $-1.0$ | $1.0$ |
| $[Fe(CN)_6]^{4-}$ | $-0.05^d$ | | | |
| $[Ru(NH_3)_6]^{2+}$ | $-0.92^e$ | $+0.70$ | $0.28$ | $1.7$ |
| $[Ru(CN)_6]^{4-}$ | $-0.22^f$ | | | |
| $[Ir(NH_3)^6]^{3+}$ | $-1.51^g$ | $+1.77$ | $0.19$ | $12$ |
| $[Ir(CN)_6]^{3-}$ | $+0.26^h$ | | | |

[a] Relative to $^{57}$Fe as 1.0, *cf.* Table 2.2.
[b] Change in $|\psi(0)|^2$ between $d^6$ and $d^5$ free ion configurations, relative to $^{57}$Fe, from data by Freeman & Ellis (1978).
[c] From Berrett, Fitzsimmons & Owusu (1968).
[d] From Kerler & Neuwirth (1962).
[e] From Potzel, Wagner, Zahn, Mössbauer & Danon (1970).
[f] From Kaindl, Potzel, Wagner, Zahn & Mössbauer (1969).
[g] From Wagner, Potzel, Wagner & Schmidtke (1974).

When all six ligands are identical, derivation of the partial isomer shift is straightforward, e.g.

$$p.i.s.(L)_{Fe} = \tfrac{1}{6}\delta(FeL_6) \tag{2.3}$$

The values for these key ligands can then be applied to data for mixed ligand complexes. Spot check calculations for complexes not used in the initial derivations usually show good agreement. Similar scales can be constructed for ruthenium(II), ruthenium(III), iridium(III), and gold(I), and are given in Tables 2.11 to 2.14. Comparisons between the scales are shown in Figure 2.10, which shows good agreement. (The slopes of the correlations are all negative because $^{57}$Fe is the only one of these isotopes for which $\delta R/R$ is negative.) The range of values for p.i.s.(CO)$_{Fe}$ may represent the synergistic bonding pattern for this ligand (Bancroft & Libbey, 1973). The best agreement with p.i.s.(CO)$_{Ru}$ is obtained when both values are derived

Table 2.10. *Partial isomer shift values for low spin iron(II)*[a]

| Ligand | p.i.s./mm s$^{-1}$ | Ligand | p.i.s./mm s$^{-1}$ |
|---|---|---|---|
| I$^-$ | 0.13 | $\tfrac{1}{2}$depb | 0.06 |
| Br$^-$ | 0.13 | $\tfrac{1}{2}$depe | 0.06 |
| Cl$^-$ | 0.10 | NCS$^-$ | 0.05 |
| H$_2$O | 0.10 | P(OPh)$_3$ | *ca.* 0.05 |
| AsPh$_3$ | >0.09 | P(OEt)$_3$ | 0.04 |
| N$_3^-$ | 0.08 | CN$^-$ | 0.01 |
| PPh$_3$ | *ca.* 0.07 | ArNC | 0.00 |
| $\tfrac{1}{2}$dppe | 0.07 | CO | < −0.01 |
| NH$_3$ | 0.07 | H$^-$ | −0.08 |
| PMe$_3$ | 0.06 | NO$^+$ | −0.20 |

[a] Relative to stainless steel. To convert to metallic iron, subtract 0.16 mm/s from the calculated isomer shift.
From Bancroft & Libbey (1973).

Table 2.11. *Partial isomer shift values for ruthenium(III)*

| Ligand | Complex | $\delta$/mm s$^{-1}$ | p.i.s./mm s$^{-1}$ |
|---|---|---|---|
| F$^-$ | [RuF$_6$]$^{3-}$ | −0.84 | −0.14 |
| Cl$^-$ | [RuCl$_6$]$^{3-}$ | −0.70 | −0.12 |
| Br$^-$ | [RuBr$_6$]$^{3-}$ | −0.79 | −0.13 |
| H$_2$O | [RuCl$_5$(H$_2$O)]$^{2-}$ | −0.71 | −0.12 |
| NCS$^-$ | [Ru(NCS)$_6$]$^{3-}$ | −0.49 | −0.08 |
| NH$_3$ | [Ru(NH$_3$)$_6$]$^{3+}$ | −0.50 | −0.08 |

Data from the compilation by Wagner & Wagner (1978).

from data for complexes containing similar ligands. Analogous variability is seen for p.i.s.(NO)$^+$.

The p.i.s. value represents the contribution of a ligand to the electron density in the valence shell of the metal ion. For ligands which are predominantly $\sigma$-donors, increasingly positive values (negative for $^{57}$Fe) indicate increased donation. The concurrence of the scales for a variety of

Table 2.12. *Partial isomer shift values for ruthenium*(II)

| Ligand | Complex | $\delta$/mm s$^{-1}$ | p.i.s./mm s$^{-1}$ |
|---|---|---|---|
| NH$_3$ | [Ru(NH$_3$)$_6$]$^{2+}$ | −0.92 | −0.15 |
| $\frac{1}{2}$ en | [Ru(en)$_3$]$^{2+}$ | −0.85 | −0.14 |
| $\frac{1}{2}$ bipy | [Ru(bipy)$_3$]$^{2+}$ | −0.54 | −0.09 |
| CN$^-$ | [Ru(CN)$_6$]$^{4-}$ | −0.23 | −0.04 |
| NO$^+$ | [Ru(CN)$_5$NO]$^{2-}$ | −0.06 | +0.13 |
| | [Ru(NH$_3$)$_5$NO]$^{3+}$ | −0.19 | +0.58 |
| Cl$^-$ | [RuCl$_5$NO]$^{2-}$ | −0.36 | −0.10 |
| Br$^-$ | [RuBr$_5$NO]$^{2-}$ | −0.47 | −0.12 |
| py | [Ru(NH$_3$)$_5$py]$^{2+}$ | −0.89 | −0.12 |
| N$_2$ | [Ru(NH$_3$)$_5$N$_2$]$^{2+}$ | −0.79 | −0.02 |
| CO | [Ru(NH$_3$)$_5$CO]$^{2+}$ | −0.54 | +0.22 |
| OH$^-$ | [Ru(NH$_3$)$_5$(OH)(NO)]$^{2+}$ | −0.15 | −0.11 |

Data from the compilation by Wagner & Wagner (1978).

Table 2.13. *Partial isomer shift values for iridium*(III)$^a$

| Ligand | Complex | $\delta$/mm s$^{-1}$ | p.i.s./mm s$^{-1}$ |
|---|---|---|---|
| Cl$^-$ | [IrCl$_6$]$^{3-}$ | −2.17 | −0.36 |
| Br$^-$ | [IrBr$_6$]$^{3-}$ | −2.23 | −0.37 |
| CN$^-$ | [Ir(CN)$_6$]$^{3-}$ | +0.26 | +0.04 |
| SCN$^-$ | [Ir(SCN)$_6$]$^{3-}$ | −1.65 | −0.28 |
| NH$_3$ | [Ir(NH$_3$)$_6$]$^{3+}$ | −1.51 | −0.25 |
| NO$_2$$^-$ | [IrCl$_2$(NO$_2$)$_4$]$^{3-}$ | −1.27 | −0.13 |
| py | [IrCl$_2$(py)$_4$]$^+$ | −1.30 | −0.13 |
| PMe$_2$Ph | [IrCl$_4$(PMe$_2$Ph)$_2$]$^-$ | −1.31 | +0.10 |
| $\frac{1}{2}$ dppe | [IrCl$_2$(dppe)$_2$]$^+$ | −0.82 | −0.02 |
| AsMePh$_2$ | [IrCl$_4$(AsMePh$_2$)$_2$]$^-$ | −1.60 | −0.02 |
| H$^-$ | [IrHCl(dppe)$_2$]$^+$ | −0.12 | +0.33 |
| O$_2$ | [Ir(O$_2$)(dppe)$_2$]$^+$ | −0.26 | −0.10 |

$^a$ The scales originally proposed by Williams, Jones & Maddock (1975) adopted an unnecessary arbitrary standard of p.i.s.(Cl$^-$) = 0. The above values have been recalculated and are absolute.
Data are from the compilation by Wagner & Wagner (1978).

metal ions confirms the intuitive view that a set of ligands can be placed in order of increasing donor power, which is largely independent of the identity of the acceptor. The observed series is $X^- < H_2O < $ N-donor $ < $ P-donor $ < CN^-$ which, perhaps surprisingly, resembles the spectrochemical series rather than the nephelauxetic series of increasing covalency. The reason for this is probably that p.i.s.(L), like the ligand field splitting energy,

Table 2.14. *Partial isomer shift values for gold*(I)

| Ligand | p.i.s./mm s$^{-1}$ | Ligand | p.i.s./mm s$^{-1}$ |
|---|---|---|---|
| Cl$^-$ | 0.92 | AsMePh$_2$ | 2.04 |
| Br$^-$ | 0.83 | PPh$_3$ | 2.59 |
| I$^-$ | 0.87 | PMePh$_2$ | 2.42 |
| N$_3^-$ | 1.30 | PMe$_2$Ph | 2.61 |
| py | 1.60 | PEt$_3$ | 2.70 |
| SMe$_2$ | 1.72 | CN$^-$ | 2.17 |
| AsPh$_3$ | 1.99 | C$_6$H$_4$NMe$_2^-$ | 3.43 |

Fig. 2.10. Comparison of partial isomer shift scales for iron(II) and (*a*) ruthenium(II), (*b*) ruthenium(III), (*c*) iridium(III), and (*d*) gold(I). The best fit lines are: p.i.s.(L)$_{Ru^{II}}$ = $-0.82$ p.i.s.(L)$_{Fe}$ $-0.04$ mm/s, ($r^2 = 0.92$); p.i.s.(L)$_{Ru^{III}}$ = $-0.74$ p.i.s.(L)$_{Fe}$ $-0.04$ mm/s, ($r^2 = 0.89$); p.i.s.(L)$_{Ir}$ = $-3.4$ p.i.s.(L)$_{Fe}$ + $0.07$ mm/s, ($r^2 = 0.89$); p.i.s.(L)$_{Au}$ = $-13$ p.i.s.(L)$_{Fe}$ + 2.6 mm/s, ($r^2 = 0.54$).

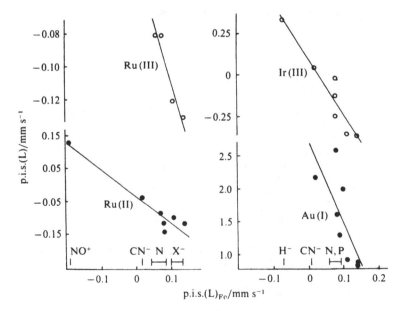

is also responsive to $\pi$-bonding effects (Bancroft *et al.*, 1970). A $\pi$-acid ligand will show an increased contribution to the isomer shift by removal from the valence shell of non-bonding $d$ electron density. Similarly, $\pi$-donor ligands will make somewhat smaller negative contributions to the isomer shift. The ligand field parameter responds in an analogous way to metal–ligand $\pi$-bonding.

### 2.5 Comparisons with other methods

Other experimental parameters which are sensitive to electron densities, and hence to bonding, are NMR chemical shifts and coupling constants, and electron binding energies. It is also possible to correlate calculated electron densities with isomer shift values (see next section).

The interpretation of the NMR parameters is difficult, because both depend on factors other than electron densities (Harris & Mann, 1978). For instance, chemical shifts depend not only on the direct shielding effect of the total electron density, but also on the electron densities on the surrounding atoms and on excitation energies. Similarly, spin–spin coupling constants depend on $s$ electron densities and on $s$ orbital participation in bonding, through the Fermi contact term, but also on excitation energies and dipolar orbital terms. Some correlation is possible between coupling constants and the Mössbauer isomer shift, as was shown in Table 2.7, but there is no good general correlation.

Better correspondence is found with binding energies estimated by X-ray photoelectron spectroscopy (XPS). The binding energy should depend principally on the total electron density on the atom concerned, as does the isomer shift. Increased donation by the ligands should therefore decrease the binding energy and increase the electron density at the nucleus. Hence, for isotopes with positive values of $\delta R/R$, a negative correlation between binding energy and isomer shift would be expected.

Several sets of data are available for tin compounds, and modest correlations are observed between the binding energy for $3d$ electrons and the isomer shift (Figures 2.11 and 2.12). The scatter probably represents the relatively poor resolution inherent in XPS measurements and the difficulties associated with charging effects; the latter are also the probable cause of differences between data from different laboratories. However, Furlani *et al.* (1981) show that the correlation is considerably improved if the XPS data are corrected for the Madelung potential of the neighbouring atoms (Figure 2.12(*b*)). None of the correlations extends to include data for tin(II) compounds.

When the data are limited to series of closely related compounds, quite good correlations are found, e.g. for $SnX_4Y_2^{2-}$ (Swartz *et al.*, 1972),

$R_2Sn(oxin)_2$ (Barber, Swift, Cunningham & Frazer, 1970), $SnCl_4L_2$ and $SnBr_4L_2$ (Grutsch, Zeller & Fehlner, 1973), or $Et_2SnX_2$ and $Ph_4Sn_2(O_2CR)_2$ (Baldy, Limouzin, Maire & Mondou, 1976). The slope of the correlation varies widely, and for the last set of compounds is effectively zero.

Fig. 2.11. Correlation of $^{119}Sn$ isomer shifts (relative to $SnO_2$) with binding energies of $3d_{\frac{5}{2}}$ electrons. Symbol codes: $+$, $SnX_4Y_2{}^{2-}$ (X, Y = halogen); $\blacktriangle$, $SnCl_3$-metal complexes; $\bigcirc$, complexes $SnBr_4L_2$; $\bullet$, complexes $SnCl_4L_2$. The experimental errors are indicated on the points for $SnO_2$ and $SnCl_4F_2{}^{2-}$. (From Swartz *et al.*, 1972, and Grutsch, Zeller, & Fehlner, 1973.)

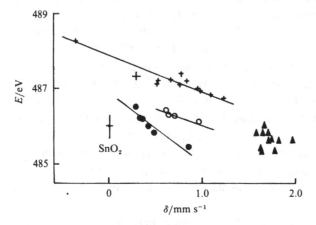

Fig. 2.12. Correlation of $^{119}Sn$ isomer shifts (relative to $SnO_2$) with electron binding energies for tin(IV) complexes: (a) $\delta$ against $E(3d_{\frac{5}{2}})$, (b) $\delta$ against $E(3d_{\frac{5}{2}}) - V(r)$, where $V(r)$ is the Madelung potential of the neighbouring atoms. Error bars are shown on one point only. (From Furlani *et al.*, 1981.)

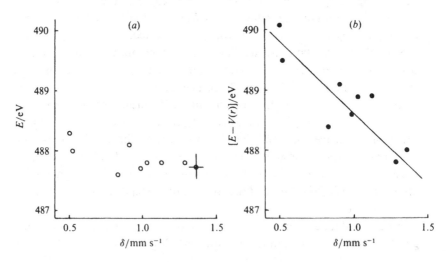

Data for low spin iron(II) complexes show a similar correlation to those for tin(IV), i.e. the binding energy decreases as the isomer shift increases (Table 2.15; Adams *et al.*, 1972). This result requires a different interpretation, however, since $\delta R/R$ is negative. The decrease in binding energy is therefore associated with a decrease in electron density at the nucleus. It is probably significant that some of the ligands involved are $\pi$-acceptors (CNAr, NO$^+$). The binding energy would be expected to decrease with increasing $\sigma$-donation, as for tin(IV), but to increase with increasing $\pi$-acceptance by the ligands, which delocalises $d$ electron density. The isomer shift, on the other hand, is expected to decrease in both cases.

### 2.6    Correlation of isomer shift with electron configuration

From the point of view of chemical bonding, it would be very convenient to be able to relate the isomer shift directly to the electron configuration of the Mössbauer atom. To this end, a knowledge is required of the dependence of the isomer shift on the population of the valence shell orbitals.

For the main group elements a relationship between isomer shift and electron populations, such as that given below, should be a good approximation.

$$\delta = An_s + Bn_s{}^2 + (Cn_s + D)n_p \qquad (2.4)$$

Where $n_s$ and $n_p$ are the populations of $s$ and $p$ electrons, the coefficient $A$ quantifies the direct effect of the valence shell $s$ and $p_{\frac{1}{2}}$ electrons and their shielding of the core electrons, and the coefficients $B$, $C$ and $D$ quantify the mutual shielding of the $s$ electrons, the shielding of $p$ electrons by $s$ electrons, and the shielding of core electrons by $p$ electrons respectively. The signs of $B$, $C$ and $D$ would therefore be expected to be opposite to that of $A$. It is usually impractical to employ so many parameters and a simpler form

Table 2.15. *XPS binding energies and Mössbauer isomer shifts for low spin iron(II) complexes*

|  | $E(2p_{\frac{3}{2}})$/eV | $E(3p)$/eV | $\delta$/mm s$^{-1}$ |
|---|---|---|---|
| [Fe(phen)$_3$]$^{2+}$ | 706.6 | 54.0 | 0.33 |
| *trans* [FeCl$_2$(CNAr)$_4$] | 707.3 |  | 0.11 |
| *trans* [Fe(SnCl$_3$)$_2$(CNAr)$_4$] | 707.7 | 54.7 | $-0.01$ |
| [Fe(SnCl$_3$)(CNAr)$_5$]$^+$ | 708.0 | 55.2 | $-0.07$ |
| [Fe(CN)$_5$(NO)]$^{2-}$ | 709.1 | 55.9 | $-0.25$ |

From Adams *et al.* (1972).

is often used:

$$\delta = an_s + bn_p \tag{2.5}$$

To evaluate the coefficients it is necessary to know the isomer shift of at least two systems for which the electron configurations are also known, which is a matter of some difficulty (Parish, 1982*b*).

For iodine and tellurium in their lower oxidation states, the bonding seems frequently to involve mainly the $5p$ orbitals, so that $b$ can be evaluated fairly readily. Data for alkali iodides ($I^-$; $5s^2 5p^6$) and molecular iodine ($I_2$; $5s^2 5p^5$) give $b_{129} = 1.50$ mm/s and $b_{127} = -0.56$ mm/s for $^{129}I$ and $^{127}I$ respectively (Bukshpan, Goldstein & Sonnino, 1968; Pasternak, Simopoulos & Hazony, 1965; Perlow & Perlow, 1966; Ehrlich & Kaplan, 1971).

Two methods have been used to derive values for $a$. Quadrupole splitting data allow estimates to be made of the $5p$ population (see later), and Hartree–Fock calculations provide estimates of the shielding factors, giving $a_{129} = -9.2$ mm/s and $a_{127} = +3.1$ mm/s (Perlow & Perlow, 1966). The same values were obtained from data for $IO_4^-$ and $IO_3^-$, assuming pure $p$ bonding for the latter ($n_s = 2$), and the same ionic character in the bonding of both species (Pasternak & Sonnino, 1968). In view of the unrealistic nature of these assumptions (Jones, 1975), the agreement is somewhat surprising.

For tellurium, nuclear quadrupole resonance (NQR) data give an estimate of bond ionicities for $TeX_6^{2-}$ (X = Cl, Br, I) which, together with the datum for CaTe (assumed to contain the closed shell $Te^{2-}$ ion), gave $b_{125} = 0.45$ mm/s (Cheyne, Johnston & Jones, 1972). The same value was obtained from the observed linear relationship for the isomer shift of isoelectronic compounds of iodine and tellurium. An approximate value for $a_{125}$ of $-2.7$ mm/s may be obtained from the same ratio (Parish, 1982*b*).

The cases of antimony and tin are more difficult, because both $s$ and $p$ orbitals are involved in the bonding, and there are substantial variations in hybridisation from compound to compound. Several estimates are based on the assumption that $KSbF_5$ and $K_2SnF_6$ correspond to the electron configurations $5s^0 5p^0$, while InSb and $\alpha$-Sn correspond to $5s^1 p^3$, none of which is very realistic (Parish, 1982*b*). The best available calibration appears to be that based on Wolfsberg–Helmholtz molecular orbital calculations for a variety of antimony compounds (Baltrunas, Ionov, Aleksandrov & Makarov, 1973); the resulting orbital populations were fitted to isomer shift data to give $a_{121} = -14.4$ mm/s and $b_{121} = 0.61$ mm/s (Parish, 1982*b*). Values for $^{119}Sn$ can be similarly derived from molecular orbital calculations for tin(IV) compounds (Greenwood, Perkins & Wall, 1967), and from the observed correlation of the isomer shift for isoelectronic tin(IV) and antimony(V) compounds (Ruby, Kalvius, Beard & Snyder, 1967), giving $a_{119} = 2.7$ mm/s and $b_{119} = -0.15$ mm/s.

In view of the assumptions and estimates involved in the derivation of the values quoted above, it is likely that they have a considerable uncertainty, probably of around 10 to 20%.

The electron configurations can also be combined with electron density plots of the type given in Figure 2.1(*b*) and with the isomer shift data; least squares fitting to Equation (2.4) then gives (Parish, 1982*b*),

$$\delta(^{121}\text{Sb})/\text{mm s}^{-1} = -19.2n_s + 1.56n_s{}^2 + 0.28n_p + 0.95n_s n_p \qquad (2.6)$$

$$\delta(^{119}\text{Sn})/\text{mm s}^{-1} = 4.08n_s - 0.38n_s{}^2 - 0.07n_p - 0.21n_s n_p \qquad (2.7)$$

With transition elements the situation is more complicated. In addition to the non-bonding $(n-1)d$ population, which changes with oxidation state, the $(n-1)d$, $ns$, and $np$ orbitals are populated by bonding interactions with the ligands. The direct contribution of the $s$ electrons will be affected by the shielding of $d$ and $p$ electrons and this, together with mutual shielding interactions, would give a very complex dependence of isomer shift on electron configuration. The only practical way to quantify this dependence would be by the combination of isomer shift data with detailed molecular orbital calculations for a range of species. Many calculations have been performed for iron compounds, usually with the aim of deriving the electron density at the nucleus and hence $\delta R/R$. Unfortunately for the present purpose, the calculated configurations cover only narrow ranges of $s$ populations (e.g. $0.116 < n_s < 0.203$; Reschke, Trautwein & Desclaux, 1977), and $p$ orbitals are often neglected. Owing to the range of oxidation states available, greater variation is found in the $d$ population; fits to the widest range of data available (Reschke *et al.*, 1977, and references therein) suggest that for high spin systems a unit change in $n_d$ corresponds to a change in isomer shift of about 2 mm/s. This figure is probably only reliable to about $\pm 0.5$ mm/s, and corresponds to the free ion values for small $n_s$ (Gütlich, Link & Trautwein, 1978; Walker, Wertheim & Jaccarino, 1961).

Gold is intermediate between transition and non-transition metal behaviour, since gold(I) has a closed shell configuration, $5d^{10}$. Based on the isomer shift for gold metal (assumed to be $5d^{10}6s^1$) and the alkali metal aurides (Au$^-$, $5d^{10}6s^2$), it has been suggested that unit increase in $n_s$ corresponds to an increase in isomer shift of about 8 mm/s (Sham, Watson & Perlman, 1981).

## 3 Quadrupole splitting as a probe for bonding
### 3.1 *The quadrupole interaction and electric field gradients*

A quadrupole split Mössbauer spectrum results from a nucleus with an electric quadrupole moment which experiences an electric field gradient (EFG). A quadrupole moment occurs when the nucleus has spin greater than one-half, and all Mössbauer nuclei possess a quadrupole moment in either the ground or excited state or both, and are therefore liable to show quadrupole splitting. The spectra are characterised by the quadrupole coupling constant, which is the product of $V_{zz}$ the principal component of the EFG (often designated as $eq$) and $eQ$ the electric quadrupole moment of the nucleus, where $e$ is the charge on the proton. This quadrupole coupling constant is the same as that derived from pure nuclear quadrupole resonance (NQR) spectra. However, the Mössbauer technique has the dual advantages of being easier in practice, and of having a wider range of isotopes available, since in cases where the ground state has $Q=0$ (and is therefore NQR silent) the excited state always has a quadrupole moment. Additionally, for nuclear spin $I > \frac{3}{2}$ the Mössbauer spectrum gives the sign of the quadrupole coupling constant directly; in other cases the sign can often be obtained from spectra obtained with an applied magnetic field. For Mössbauer transitions between nuclear states with spin $\frac{1}{2}$ and $\frac{3}{2}$ (e.g. $^{57}$Fe, $^{119}$Sn, $^{191}$Ir, $^{197}$Au), the spectra are simple doublets, which are split by an amount $\Delta = eQV_{zz}/2$, which is known as the quadrupole splitting and is one-half the quadrupole coupling constant for the $I = \frac{3}{2}$ state. When the spin is greater the spectra are more complex and data are presented as the quadrupole coupling constant for the ground state (or the excited state, when the ground state has $Q=0$).

An EFG results whenever the distribution of charge about the nucleus has less than cubic symmetry. The EFG can be specified by its principal component:

$$V_{zz} = \frac{\partial^2 V}{\partial z^2} = \frac{1}{4\pi\varepsilon_0} \sum_i q_i r_i^{-3} (3\cos^2\theta_i - 1) \tag{2.8}$$

where the summation is over all charges $q_i$ with polar coordinates $r_i$ and $\theta_i$. The charges concerned may be external to the Mössbauer atom (lattice charges) giving a contribution $(V_{zz})_{lat}$ or may be due to the electrons of the bond system or to the non-bonding electrons, giving contributions $(V_{zz})_b$ and $(V_{zz})_{nb}$ respectively. The last two are often treated together as the valence shell contribution $(V_{zz})_{val}$. Because of the inverse cubic dependence on distance from the nucleus, these effects become increasingly important in the order $(V_{zz})_{lat} < (V_{zz})_b < (V_{zz})_{nb}$. Except in ionic systems, the lattice contribution is usually negligible. The EFG at the nucleus is also affected by shielding and antishielding effects due to polarisation of the core electrons (Sternheimer, 1956). These factors are implicitly included in all the discussions below.

## 3.2 The interpretation of quadrupole splitting

In systems other than closed shell ionic lattices, $(V_{zz})_{lat}$ can normally be ignored. The magnitude of $(V_{zz})_b$ depends on the nature and arrangement of the ligands, and is discussed in some detail below. When non-bonding electrons are present their effects are nearly always evident, regardless of the ligands.

### 3.2.1 EFG due to non-bonding electrons

For $(V_{zz})_{nb}$ to be non-zero, the population of the valence shell $p$ orbitals (of main group elements) or $d$ orbitals (in the case of transition metals) must be non-equivalent. Thus for tin(II) or antimony(III), the valence shell contains a single lone pair, which normally contributes significantly to the EFG. Only when the metal atom occupies a site with rigorous cubic symmetry, as in $CsSnBr_3$ or $Cs_3SbCl_6$, is $(V_{zz})_{nb}$ zero. In these cases the lone pair has a high $5s$ character.

When the symmetry is lower, the lone pair is often stereochemically active, and is usually directed along the major symmetry axis of the system. Since this axis is also normally taken as the $z$ axis, this means that the $5p_z$ population is higher than that of $5p_x$ or $5p_y$. Such concentration of negative charge on the $z$ axis gives substantial quadrupole splitting, with $(V_{zz})_{nb} < 0$ (Table 2.16; for both $^{119}Sn$ and $^{121}Sb$, $Q < 0$: positive values for $eQV_{zz}$ therefore indicate that $V_{zz}$ is negative).

A similar situation arises when two lone pairs are present, as for tellurium(IV) and iodine(III). In these cases, the bonding seems to involve

Table 2.16. *Quadrupole splitting data for tin(II) and antimony(III)*

|  | $\Delta/mm\ s^{-1}$ |  | $\Delta/mm\ s^{-1}$ |
|---|---|---|---|
| SnS | 1.0[a] | SbCl$_3$ | 12.2,[b] 13.9[c] |
| SnC$_2$O$_4$ | 1.8[b] | Et$_4$N[SbCl$_4$] | 8.9,[c] 10.6[d] |
| K$_2$[Sn(C$_2$O$_4$)$_2$] . H$_2$O | 1.97[e] | (NH$_4$)$_2$[SbCl$_5$] | 11.2[f] |
| Sn(O$_2$CCH$_3$)$_2$ | 1.56[e] | SbPh$_3$ | 16.2,[g] 17.5,[h] 17.0[i] |
| NaSnF$_3$ | 1.84[e] | SbMe$_3$ | 16.3[i] |

[a] From Gibb, Goodman & Greenwood (1970).
[b] From Birchall, Della Valle, Martineau & Milne (1971).
[c] From Donaldson, Southern & Tricker (1972).
[d] From Birchall, Ballard, Milne & Moffatt (1974).
[e] From Donaldson, Filmore & Tricker (1971).
[f] From Birchall & Della Valle (1971).
[g] From McAuliffe, Niven & Parish (1977b).
[h] From Long, Stevens, Tullbane & Bowen (1970).
[i] From Brill, Parris, Long & Bowen (1973).

principally the $p$ orbitals. One lone pair has high $5s$ character and the other high $5p$ character. (Alternatively and equivalently the two lone pairs could be regarded as occupying two $sp$ hybrid orbitals.) The EFG is again dominated by the $5p$ electron density, and $(V_{zz})_b$ is negative. For instance, iodine(III) usually adopts square-planar coordination as in $I_2Cl_6$ or $KICl_4$, with the lone pair electron density perpendicular to the plane. Since $Q_{127}$ (the quadrupole moment of the $^{127}I$ nucleus) is negative, the quadrupole coupling constant is positive: $+3060$ and $+3094$ MHz respectively (Pasternak & Sonnino, 1968; Perlow & Perlow, 1966). Similar considerations apply to tellurium(II) and tellurium(IV) compounds (Jones, Schultz, McWhinnie & Dance, 1976). It should be noted that it is customary to present data for iodine compounds as $eQ_{127}V_{zz}/$MHz, in order to facilitate comparison with NQR data. Conversion factors are $eQ_{127}V_{zz}/$MHz$ = 46.46eQ_{127}V_{zz}/$mm s$^{-1} = 32.58eQ_{129}V_{zz}/$mm s$^{-1}$.

Cases in which there are three lone pairs, as in iodine(I) compounds, are better considered as being derived from closed shell configurations by the loss of electron density from one $p$ orbital to the neighbouring atom. The quadrupole coupling constant then reflects the extent of electron transfer, and follows the electronegativity of the neighbour (Table 2.17).

In transition metal compounds $(V_{zz})_{nb}$ is zero for high spin $d^5$, low spin $d^6$, and $d^{10}$ configurations. In all other cases, imbalance in the $d$ orbital populations .gives rise to contributions to the EFG which may be substantial. For instance, octahedral high spin iron(II) compounds have quadrupole splitting values of 2–4 mm/s, representing principally the $(V_{zz})_{nb}$ contribution of the sixth $t_{2g}$ electron. The full contribution can only be manifested if there are deviations from exact cubic symmetry since, if the $t_{2g}$ orbitals are completely degenerate, this electron would be delocalised over the three orbitals $d_{xy}$, $d_{xz}$ and $d_{yz}$, and $(V_{zz})_{nb}$ would average to zero. Only if the symmetry is lowered (as the Jahn–Teller theorem requires) will this electron become localised and make a contribution to the EFG. Thus a small axial compression will remove the degeneracy of the $t_{2g}$ set, causing

Table 2.17. *Quadrupole coupling constants for iodine(I) species*[a]

| ICl | IBr | $I_2$ | HI |
|-----|-----|-------|-----|
| −3000 | −2982 | −2238 | −1640 |

[a] The data quoted are values of $eQ_{127}V_{zz}/$MHz, and are from the compilation by Bancroft & Platt (1972).

$d_{xy}$ to be depressed in energy (Figure 2.13); the resulting localisation of the paired electron in this orbital gives $(V_{zz})_{nb} > 0$ and quadrupole splitting values of up to around 4 mm/s. Since the relative contributions to $(V_{zz})_{nb}$ of an electron in $d_{xy}$, $d_{xz}$, or $d_{yz}$ are $-2:1:1$ (see Table 2.23), an opposite distortion which localised the electron in $d_{xz}$ and $d_{yz}$ orbitals, would make $(V_{zz})_{nb} < 0$, with half the previous magnitude. Similar effects occur for tetrahedral complexes. The additional ligand field splitting energies are often small, allowing thermal excitation and giving considerable temperature dependence of the quadrupole splitting. Analogous though less marked effects are seen with low spin iron(III); in this case $(V_{zz})_{nb}$ is reduced by covalent delocalisation of the $d$ electron density. In principle, the extent of distortion, the degree of covalency, and the spin–orbit coupling constant can be evaluated from detailed analysis of the temperature dependence of the quadrupole splitting (Gibb, 1968; Golding, 1976; Ingalls, 1964). In practice, however, these variables are highly correlated, and it is difficult to obtain unique solutions. Similar problems are encountered in the analogous analysis of magnetic susceptibility data.

Significant $(V_{zz})_{nb}$ contributions to the EFG also arise in metal ions with $d^8$ configurations, such as nickel(II), iridium(I) and gold(III). In these cases, coordination is often square planar, and excess electron density lies above and below the coordination plane. This case is discussed further below.

### 3.2.2 *EFG due to bonding electrons*

The effects of asymmetry in the distribution of bonding electrons $(V_{zz})_b$ are superimposed on $(V_{zz})_{nb}$. In cases where the Mössbauer atom may be treated as a complexed cation, these two contributions are usually

Fig. 2.13. Crystal field splitting diagrams showing the effects of tetragonal compressive distortions on (*a*) octahedral and (*b*) tetrahedral $d^6$ systems.

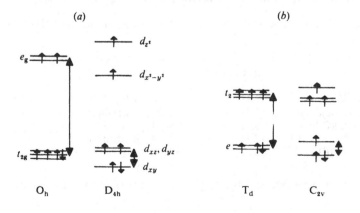

opposite in sign. For instance in the trihalogenostannates(II) $SnX_3^-$, the lone pair lies on one side of the tin atom and the three halide ligands on the other, in a pyramidal arrangement. The EFG due to the lone pair is offset by donation from the halide ions, which increases in the order $X = F < Cl < Br < I$; this is also the order of decreasing quadrupole splitting (Clark, Maresca & Smith, 1970). The same trend is seen in the isoelectronic antimony(III) halides (Bancroft & Platt, 1972).

In transition metal complexes, non-bonding electrons are often not stereochemically active, but similar effects occur. For instance, with high spin iron(II), removal of the $t_{2g}$ degeneracy occurs with *trans* $FeA_2B_4$ coordination. The maximum quadrupole splitting is obtained when $d_{xy}$ lies lowest in energy; $(V_{zz})_{nb}$ is then large and positive. This requires A to be the stronger field ligand; that is, ligand A must donate more electron density to the bonding orbitals along the $z$ axis than B does to those in the $xy$ plane; $(V_{zz})_b$ is therefore negative.

In square-planar $d^8$ systems, $(V_{zz})_{nb}$ originates from the pair of electrons in the $d_{z^2}$ orbital perpendicular to the coordination plane, and is therefore negative. The ligands provide electron density in the $xy$ plane, so that $(V_{zz})_b$ is positive. Thus, for $[Ni(niox)_2]$ (niox is the monoanion of 1,2-cyclohexanedionedioxime) the quadrupole coupling constant is negative showing that $|(V_{zz})_{nb}| > |(V_{zz})_b|$ (Dale, Dickinson & Parish, 1979). However, when donation is increased by the use of stronger ligands, as in $K_2[Ni(CN)_4]$, the quadrupole splitting cannot be resolved, indicating that $(V_{zz})_{nb}$ and $(V_{zz})_b$ are now nearly equal. In systems where the metal–ligand bonds are highly covalent, such as gold(III) or iridium(I), $(V_{zz})_b$ is usually the dominant factor. This can be deduced from the trend in values of quadrupole splitting as the ligands are systematically changed (Table 2.18), and has been confirmed by direct measurement in the case of $K[Au(CN)_4]$ (Bartunik, Holzapfel & Mössbauer, 1970).

In many cases non-bonding electrons are not present or, if present, do not

Table 2.18. *Quadrupole splitting data for iridium*(I) *and gold*(III) *complexes*

| | $\Delta/mm\ s^{-1}$ | | $\Delta/mm\ s^{-1}$ |
|---|---|---|---|
| $[IrCl(CO)_3]$ | 2.11 | $Bu_4N[AuBr_4]$ | 1.50 |
| $Ph_4As[IrBr_2(CO)_2]$ | 5.32 | $[AuBr_2(S_2CNBu_2)]$ | 2.20 |
| $[IrBr(CO)(PPh_3)_2]$ | 6.92 | $[AuMe_2(S_2CNBu_2)]$ | 5.11 |
| $[Ir(CO)(PPh_3)_3]PF_6$ | 8.31 | $[AuMe_2Br(PPh_3)]$ | 6.30 |
| $[Ir(dppe)_2]PF_6$ | 8.8 | $K[Au(CN)_4]$ | 6.9 |

Data from the compilations by Wagner & Wagner (1978) and Parish (1982a).

contribute to the EFG. Quadrupole splitting is then due entirely to $(V_{zz})_b$ and, as implied above, can arise in two essentially different ways. It is always necessary that the symmetry of the environment of the Mössbauer atom be less than cubic, but this may be achieved either by an arrangement of identical ligands which has low symmetry or by the presence of non-equivalent ligands in regular octahedral or tetrahedral geometry. In the latter case, $(V_{zz})_b$ represents the difference in bonding characteristics of the ligands. If only two different ligands are involved, $(V_{zz})_b$ increases in magnitude as the difference in donor power increases. Table 2.19 shows data for a low spin $d^6$ case, $[Ir(NH_3)_5L]^{2+}$, where the quadrupole splitting increases in the order $L = NCS^- < NO_2^- < O_2CCH_3^- < O_2CH^- < Cl^- < I^-$ (the small value for $[Ir(NH_3)_6]I_3$ results from a lattice contribution). Thus, as L becomes progressively less like $NH_3$ in donor power, $(V_{zz})_b$ and the quadrupole splitting increase. Similar effects are seen for tetrahedral tin(IV) compounds (Table 2.20). (Although these compounds are generally

Table 2.19. *Quadrupole splitting data for iridium(III) ammines*

|  | $\Delta/mm\ s^{-1}$ |
|---|---|
| $[Ir(NH_3)_6]I_3$ | 0.14 |
| $[Ir(NH_3)_5(NCS)](ClO_4)_2$ | 0.31 |
| $[Ir(NH_3)_5(NO_2)]Cl_2$ | 0.70 |
| $[Ir(NH_3)_5(O_2CCH_3)](ClO_4)_2$ | 1.14 |
| $[Ir(NH_3)_5(O_2CH)](ClO_4)_2$ | 1.60 |
| $[Ir(NH_3)_5Cl]Cl_2$ | 2.06 |
| $[Ir(NH_3)_5I]Cl_2$ | 2.72 |

Data from the compilation by Wagner & Wagner (1978).

Table 2.20. *Quadrupole splitting data for tetrahedral organotin(IV) compounds*

|  | $\Delta/mm\ s^{-1}$ |
|---|---|
| $Me_4Sn$ | 0.00 |
| $Me_3SnC_6H_5$ | 0.00 |
| $Me_3SnC_6Cl_5$ | 1.09 |
| $Me_3SnC\equiv CPh$ | 1.17 |
| $Me_3SnCCl=CCl_2$ | 1.24 |
| $Me_3SnC_6F_5$ | 1.31 |
| $Me_3SnCF_3$ | 1.38 |

Data from Parish & Platt (1969).

considered as covalent molecules, they can equally well be treated as tetrahedral complexes of $Sn^{4+}$ with carbanion ligands.) In each of these cases it is intuitively obvious that the substituents are poorer donors than the constant ligands, and $(V_{zz})_b$ should be positive. Confirmation could be achieved by determining the sign of the EFG.

Such confirmation is more necessary when unconventional ligands are involved, such as the hydride ion. The low spin iron(II) complex *trans* $[FeH_2(depb)_2]$ shows a quadrupole splitting of 1.84 mm/s, suggesting that $H^-$ is an appreciably different donor from the phosphine (Bancroft *et al.*, 1970). However, the corresponding dichloro complex gives a value of 1.13 mm/s, and the analogous complex *trans* $[FeCl_2(depe)_2]$ gives a value of 1.29 mm/s. While it is expected that $Cl^-$ would be rather different from a P-donor, it is difficult to compare $Cl^-$ and $H^-$. The observation that *trans* $[FeCl(H)(depe)_2]$ has an unresolved quadrupole splitting suggests that the two differences ($Cl^-$ compared with P, and $H^-$ compared with P) are opposite in sign, indicating that the order of donation to iron(II) is $Cl^- < P < H^-$. This was confirmed by a measurement of the sign of the EFG for $[FeCl_2(depe)_2]$, which showed that $(V_{zz})_b$ is positive (Bancroft *et al.*, 1971). The data of Table 2.21 for $[FeH(L)(depe)_2]$ can consequently be interpreted as showing that the donation from L increases as the quadrupole splitting increases; in confirmation, $(V_{zz})_b$ was found to be negative for $L = CNC_6H_4OMe$.

The observed lack of a quadrupole splitting for $[FeCl(H)(depe)_2]$ suggests that $(V_{zz})_b$ might be an additive function of the ligands. It should therefore be possible to assign scales of partial quadrupole splitting parameters to the ligands, and this is illustrated below. Such scales allow

Table 2.21. *Quadrupole splittings for trans* $[Fe(H)(L)(depe)_2]BPh_4$

| L | $\Delta/$mm s$^{-1}$ |
| --- | --- |
| I | *ca.* 0 |
| Cl | *ca.* 0 |
| $N_2$ | 0.33 |
| $CH_3CN$ | 0.46 |
| PhCN | 0.58 |
| $(PhO)_3P$ | 0.72 |
| $(MeO)_3P$ | 0.90 |
| $Bu^tNC$ | 1.13 |
| $p\text{-}MeOC_6H_4NC$ | 1.14 |

Data from Bancroft, Mays, Prater & Stefanini (1970).

further comparisons to be made of the relative bonding characteristics of a variety of ligands.

When the basic coordination geometry is neither octahedral nor tetrahedral, an EFG is expected whether the ligands differ or not. Examples are known of two-, three- and five-coordination, such as $[Au(CN)_2]^-$, $[Au(PPh_3)_3]^+$, and $(SnCl_5)^-$, in addition to the square-planar complexes mentioned above. In all of these cases $(V_{zz})_{nb}$ is zero and the EFG arises principally from the asymmetric distribution of bonding electron density. The situation is well illustrated by the linear two-coordinate complexes of $Au^+$. The metal ion has $sp$ hybridisation, so that the $s$ and $p_z$ orbitals are populated by donation from the ligands. Since $p_x$ and $p_y$ remain empty, a substantial EFG is created, which increases as the ligands become better donors. The quadrupole splitting therefore increases as the ligand changes from halide ions, through N- and S-donors, to P- and C-donors (Table 2.22). The implied negative sign for $(V_{zz})_b$ has been confirmed for $K[Au(CN)_2]$ (Prosser *et al.*, 1975). The similarity of this order of donor power to that for the very different acceptor Fe(II) is striking. Again, there appears to be a rough additivity of ligand contribution: the quadrupole splittings observed for $[AuCl(PPh_3)]$, 7.43 mm/s, and $[AuCl(py)]$, 6.75 mm/s, for instance, are close to the averages of the values for the appropriate parent complexes.

### 3.3 Semi-quantitative treatments of the electric field gradient

Lone pair effects which give rise to $(V_{zz})_{nb}$ are difficult to quantify, especially for the main group elements for which changes in hybridisation

Table 2.22. *Quadrupole splittings for gold*(I) *complexes*, $[AuL_2]^+$

| L | $\Delta/\text{mm s}^{-1}$ |
|---|---|
| $I^-$ | 5.75 |
| $Br^-$ | 6.35 |
| $Cl^-$ | 6.28 |
| $S_2O_3^{2-}$ | 7.01 |
| py | 7.32 |
| $SMe_2$ | 7.56 |
| $PPh_3$ | 9.53 |
| $PEtPh_2$ | 9.49 |
| $PEt_2Ph$ | 9.77 |
| $PEt_3$ | 10.18 |
| $CN^-$ | 10.10 |
| $-C_6H_4NMe_2{}^-$ | 12.01 |

Data from the compilation by Parish (1982*a*).

affect the $p$ character of the lone pair. The effects due to the bonding electrons are somewhat easier to treat, and three related approaches have been adopted. Before these are discussed, it is necessary to examine the EFG tensor in a little more detail.

### 3.3.1    The EFG tensor

For a set of charges $q_i$ with polar coordinates $r_i$, $\theta_i$, and $\phi_i$, the principal component of the EFG has already been given as

$$V_{zz} = \frac{1}{4\pi\varepsilon_0} \sum_i q_i r_i^{-3}(3\cos^2\theta_i - 1) \qquad (2.9)$$

Eight other components can be specified, but by a suitable choice of axes the off-diagonal elements ($V_{xy}$, etc.) can usually be made to vanish. The remaining diagonal elements are then given by

$$V_{yy} = \frac{1}{4\pi\varepsilon_0} \sum_i q_i r_i^{-3}(3\sin^2\theta_i \sin^2\phi_i - 1) \qquad (2.10)$$

$$V_{xx} = \frac{1}{4\pi\varepsilon_0} \sum_i q_i r_i^{-3}(3\sin^2\theta_i \cos^2\phi_i - 1) \qquad (2.11)$$

When the EFG has axial symmetry $V_{xx} = V_{yy} = -\frac{1}{2}V_{zz}$, which is a result of the Laplace condition that $V_{xx} + V_{yy} + V_{zz} = 0$. In this situation the separation of the nuclear substates and hence the magnitude of the quadrupole splitting depend only on $eQV_{zz}$. For non-axial symmetry another parameter is required to specify the EFG. This is the asymmetry parameter $\eta$, given by

$$\eta = (V_{xx} - V_{yy})/V_{zz} \qquad (2.12)$$

When the Mössbauer spectrum is a simple doublet, the quadrupole splitting $\Delta$ is then given by

$$\Delta = \frac{1}{2}eQV_{zz}(1 + \eta^2/3)^{\frac{1}{2}} \qquad (2.13)$$

More complex spectra allow, in principle at least, the direct determination of both $eQV_{zz}$ and $\eta$.

When the Mössbauer nucleus is surrounded by a set of localised charges, the above equations can be used to calculate the EFG directly, and this is the basis of one approximate treatment. More rigorously, the charges must be treated as electron density associated with atomic or molecular orbitals. The geometrical terms must then be replaced by the appropriate expectation values, giving EFG components such as

$$V_{zz} = \frac{-e}{4\pi\varepsilon_0} \sum_i \langle 3\cos^2\theta_i - 1\rangle\langle r_i^{-3}\rangle \qquad (2.14)$$

The values of these terms for $p$ and $d$ electrons are shown in Table 2.23. It is usually necessary to consider only the $p$ or the $d$ orbitals, so that the $\langle r^{-3} \rangle$ term is the same for all electrons.

### 3.3.2    The point charge and hybridisation methods

The quadrupole coupling constant depends not only on the differences in the bonding characteristics of the various ligands, but also on the number of each type of ligand and their relative disposition. For instance, it is observed that the quadrupole splitting for $MAB_5$ is about half that of *trans* $MA_2B_4$, but similar to that of *cis* $MA_2B_4$ (Table 2.24).

The simplest treatment of this situation is analogous to the crystal field approach for transition metal complexes. The ligands are considered to act as point charges, and each is assigned a partial quadrupole splitting parameter p.q.s.(L), which is assumed to be independent of the other ligands present, and to have a constant value; p.q.s.(L) thus corresponds to an effective value of $q_i r_i^{-3}$, and different values may apply for different coordination numbers. It is useful to choose parameters such that the value of the quadrupole splitting can be calculated directly:

$$\Delta = \sum_i \text{p.q.s.}(L_i)(3\cos^2\theta_i - 1) \tag{2.15}$$

that is

$$\text{p.q.s.}(L_i) = \frac{e|Q|q_i r_i^{-3}}{8\pi\varepsilon_0} \tag{2.16}$$

(It will be evident that the sign of the quadrupole splitting given by this expression is actually that of $V_{zz}$. This has the effect that all scales of partial quadrupole splitting values are on a common basis, regardless of the sign of $Q$. The parameters also take implicit account of Sternheimer antishielding.)

For a regular octahedron, $\theta_i$ will be $0°$, $90°$, or $180°$, and evaluation of the above expression leads directly to

$$\Delta = 2\,\text{p.q.s.}(A) - 2\,\text{p.q.s.}(B) \quad \text{for } MAB_5 \tag{2.17}$$

$$\Delta = 4\,\text{p.q.s.}(A) - 4\,\text{p.q.s.}(B) \quad \text{for } trans\ MA_2B_4 \tag{2.18}$$

$$\Delta = 2\,\text{p.q.s.}(B) - 2\,\text{p.q.s.}(A) \quad \text{for } cis\ MA_2B_4 \tag{2.19}$$

(In the last case, the $z$ axis is taken to coincide with the B–M–B direction, since this adventitiously leads to $\eta = 0$.) The calculated ratios of quadrupole

Table 2.23. *Expectation values of angular functions for p and d orbitals*

| Orbital | $p_x$ | $p_y$ | $p_z$ | $d_{xy}$ | $d_{xz}$ | $d_{yz}$ | $d_{x^2-y^2}$ | $d_{z^2}$ |
|---|---|---|---|---|---|---|---|---|
| $\langle 3\cos^2\theta - 1 \rangle$ | $-\frac{2}{5}$ | $-\frac{2}{5}$ | $+\frac{4}{5}$ | $-\frac{4}{7}$ | $+\frac{2}{7}$ | $+\frac{2}{7}$ | $-\frac{4}{7}$ | $+\frac{4}{7}$ |

Table 2.24. Quadrupole splittings for complexes of the types $MAB_5$, cis $MA_2B_4$ and trans $MA_2B_4$

| $MAB_5$ | $\Delta/mm\ s^{-1}$ | cis $MA_2B_4$ | $\Delta/mm\ s^{-1}$ | trans $MA_2B_4$ | $\Delta/mm\ s^{-1}$ |
|---|---|---|---|---|---|
| $[FeCl(CNAr)_5]ClO_4$ | 0.73 | $[FeCl_2(CNAr)_4]$ | 0.78 | $[FeCl_2(CNAr)_4]$ | 1.55 |
| $[Fe(CN)(CNEt)_5]ClO_4$ | 0.17 | $[Fe(CN)_2(CNEt)_4]$ | 0.29 | $[Fe(CN)_2(CNEt)_4]$ | 0.59 |
| | | $[IrCl_2(py)_4]Cl$ | 1.50 | $[IrCl_2(py)_4]Cl$ | 2.60 |
| | | $pyH[Ir(py)_2Cl_4]$ | 1.44 | $pyH[Ir(py)_2Cl_4]$ | 3.16 |
| $(Me_4N)_2[Sn(Et)Cl_5]$ | 1.94 | | | $(Me_4N)_2[SnEt_2Cl_4]$ | 3.99 |
| $(pyH)_2[Sn(Ph)Cl_5]$ | 1.92 | | | $(pyH)_2[SbPh_2Cl_4]$ | 3.80 |

Data from Bancroft et al. (1970); Berrett & Fitzsimmons (1967); Wagner & Wagner (1978); Parish (1984a).

splitting thus agree with those observed. Simple expressions also result for four-coordinate systems with basic tetrahedral geometry. To allow treatment of other geometries, and to make comparisons of the donor capacities of various ligands, it is necessary to evaluate the parameters p.q.s.(L); this is illustrated below, after consideration of the hybridisation approach.

The ligands may be viewed as populating the valence orbitals by dative bonding, in which case the EFG results from non-equivalent donation to the various $p$ or $d$ orbitals. The extent to which each ligand populates a particular orbital depends on the donor capacity of the ligand, and also on the $p$ (or $d$) character of the hybrid orbital on the acceptor (Clark, Maddock & Platt, 1972). In an octahedral complex, the bonds have one-half $p$ character and one-third $d$ character. A ligand of donor power $\sigma$, lying on the $z$ axis, therefore makes a contribution to the EFG which is proportional to

$$-\tfrac{1}{2}\langle 3\cos^2\theta - 1\rangle_p \langle r^{-3}\rangle_p \sigma - \tfrac{1}{3}\langle 3\cos^2\theta - 1\rangle_d \langle r^{-3}\rangle_d \sigma$$
$$= -\tfrac{2}{5}\langle r^{-3}\rangle_p \sigma - \tfrac{4}{21}\langle r^{-3}\rangle_d \sigma \quad (2.20)$$

while a ligand on the $x$ or $y$ axis contributes

$$+\tfrac{1}{5}\langle r^{-3}\rangle_p \sigma + \tfrac{2}{21}\langle r^{-3}\rangle_d \sigma \quad (2.21)$$

this treatment gives the same result as the simpler one if

$$\text{p.q.s.(L)} = \tfrac{1}{5}\langle r^{-3}\rangle_p \sigma - \tfrac{2}{21}\langle r^{-3}\rangle_d \sigma \quad (2.22)$$

Similar treatments apply to other geometries. It is expected that $\langle r^{-3}\rangle_p$ and $\langle r^{-3}\rangle_d$ will have very different values; $\langle r^{-3}\rangle_p$ can normally be neglected for transition metals, and $\langle r^{-3}\rangle_d$ for main group elements.

It is usually found that the two methods lead to analogous expressions. It is therefore convenient to use the point charge treatment, which appears also to be able to take account of distortions from regular geometry (Parish & Johnson, 1971; Sham & Bancroft, 1975). The hybridisation method has the virtue of emphasising that different scales of partial quadrupole splitting values are needed for different coordination geometries.

Since the EFG is a function of two (or more) partial quadrupole splitting values, an arbitrary standard has to be chosen, which is usually p.q.s.$(Cl^-) = 0$; however, for iron(II), p.q.s.$(Cl^-)$ has been set to $-0.3$ mm/s (Bancroft *et al.*, 1970). Values for the common ligands coordinated to iron(II) and iridium(III) are shown in Table 2.25. New values have been derived for ruthenium(II), which are given in Table 2.26, which also serves to illustrate the method of calculation. The different scales correlate well (Figure 2.14), indicating the essential similarity of the bonding characteristics of the ligands, which are largely independent of the acceptor. The strongest donors are those with the most negative values, e.g. $H^-$, $CN^-$

and ArNC. Tertiary phosphines are good ligands, somewhat better than N-donors, and the halide ions are relatively poor. These trends coincide, again, with normal chemical intuition.

Similar sets of values can be derived for gold(I) (Table 2.25, Figure 2.14). In this case, however, there is no need to choose an arbitrary standard: since the complexes are linear, an EFG is present even if the ligands are identical, and the quadrupole splitting of $AuL_2$ is equal to 4 p.q.s.$(L)_{Au}$. The values show the same trends as for the other metals, with p.q.s.$(L)_{Au}$ becoming more negative as the ligands become softer.

Of the main group elements, extensive partial quadrupole splitting scales have been evaluated for tin(IV) (Table 2.27). As indicated above, different

Table 2.25. *Partial quadrupole splitting values for ligands coordinated to low spin iron*(II), *iridium*(III) *or gold*(I)

| L | p.q.s.$(L)_{Fe}$/mm s$^{-1}$ [a] | p.q.s.$(L)_{Ir}$/mm s$^{-1}$ [b] | p.q.s.$(L)_{Au}$/mm s$^{-1}$ [c] |
|---|---|---|---|
| $NO^+$ | $+0.01$ | | |
| $Br^-$ | $-0.28$ | | $-1.59$ |
| $I^-$ | $-0.29$ | $+0.33$ | $-1.44$ |
| $Cl^-$ | $-0.30$ | $0.00$ | $-1.57$ |
| $N_2$ | $-0.37$ | | |
| $N_3^-$ | $-0.38$ | | $-1.71$ |
| $NO_2^-$ | $-0.41$ | $-0.68^{d,e}$ | |
| MeCN | $-0.43$ | | |
| $H_2O$ | $-0.45$ | | |
| py | $-0.47$ | $-0.75$ | $-1.83$ |
| $NCS^-$ | $-0.51$ | | |
| $AsPh_3$ | $-0.51$ | | $-2.11$ |
| $NH_3$ | $-0.52$ | $-1.03$ | |
| $PPh_3$ | $-0.53$ | | $-2.38$ |
| CO | $-0.55$ | | $-1.90$ |
| $PMePh_2$ | $-0.58$ | | $-2.37$ |
| $PMe_2Ph$ | $-0.62^f$ | $-1.46$ | $-2.41$ |
| $\frac{1}{2}$ depe | $-0.65$ | | |
| $PMe_3$ | $-0.66$ | | |
| $\frac{1}{2}$ dppe | $-0.68$ | $-0.87$ | |
| ArNC | $-0.70$ | | |
| $CN^-$ | $-0.84$ | | $-2.50$ |
| $H^-$ | $-1.04$ | $-2.24^{d,g}$ | |

[a] From Bancroft & Libbey (1973).
[b] From Williams, Jones & Maddock (1975).
[c] Derived from the quadrupole splitting of $AuL_2{}^+$ (Parish, 1984$a$).
[d] Recalculated value to give greater consistency with the iron(II) scale.
[e] Assuming a value of $-0.70$ mm/s for the quadrupole splitting of $[Ir(NH_3)_5NO_2]^+$.
[f] Interpolated value (see [b]).
[g] Assuming a value of $-1.01$ mm/s for the quadrupole splitting of $[Ir(depe)_2HCl]^+$.

scales are required for four-, five-, and six-coordination because of the different $p$ orbital participation in the hybrid orbitals employed by the tin atom; in principle, slightly different values should also be used for the apical and equatorial positions in trigonal bipyramidal coordination. As would be expected, the four sets of values correlate well, showing that change in hybridisation has no detectable effect on the relative donor powers of the ligands. Least squares fitting gives:

$$\text{p.q.s.(L)}_{Sn}^{oct} = (0.72 \pm 0.02)\text{p.q.s.(L)}_{Sn}^{tet} - (0.01 \pm 0.02) \text{ mm/s};$$

$$r^2 = 0.991 \quad (2.23)$$

$$\text{p.q.s.(L)}_{Sn}^{tbp} = (0.95 \pm 0.03)\text{p.q.s.(L)}_{Sn}^{oct} - (0.03 \pm 0.01) \text{ mm/s};$$

$$r^2 = 0.989 \quad (2.24)$$

The hybridisation treatment suggests that the slopes should be 0.67 and 1.00 respectively, assuming no significant $d$ orbital participation. The slight discrepancies may arise from some $d$ orbital involvement, or from the neglect of the electron density in the ligand valence shell.

Only limited comparisons can be made with the transition metal data, but the same broad trends are discernible: electronegative ligands such as the halide ions and O-donor ligands give small values, and the much more covalently bonded hydride and carbanionic ligands give the most negative values. The tin data seem to show rather less discrimination between the

Table 2.26. *Derivation of the partial quadrupole splitting scale for* ruthenium(II)

| L | Complex | $\Delta$/mm s$^{-1}$ [a] | p.q.s.(L)$_{Ru}$/mm s$^{-1}$ [b] |
|---|---|---|---|
| Cl$^-$ | arbitrary standard | | 0.00 |
| NO$^+$ | $[RuCl_5(NO)]^{2-}$ | 0.21 | $+0.10$ |
| Br$^-$ | $[RuBr_5(NO)]^{2-}$ | 0.08 | $+0.06$ |
| NH$_3$ | $[Ru(NH_3)_5NO]^{3+}$ | 0.36 | $-0.08$ |
| CN$^-$ | $[Ru(CN)_5NO]^{2-}$ | 0.42 | $-0.10$ |
| NCS$^-$ | $[Ru(NCS)_5NO]^{2-}$ | 0.24 | $-0.02$ |
| CO | $[Ru(NH_3)_5CO]^{2+}$ | ca. 0.10 | ca. $-0.02$ |
| MeCN | $[Ru(NH_3)_5(NCMe)]^{2+}$ | ca. 0.05 | ca. $-0.05$ |
| N$_2$ | $[Ru(NH_3)_5N_2]^{2+}$ | 0.23 | $+0.04$ |
| NO$_2{}^-$ | $[Ru(CN)_5(NO_2)]^{4-}$ | 0.28 | $+0.04$ |

[a] The EFG is assumed to be positive in all cases, as suggested by Bancroft, Butler & Libbey (1972).
[b] Sample calculations: for $RuA_5B$, $= 2$ p.q.s.(A) $- 2$ p.q.s.(B), p.q.s.(Cl$^-$) is assumed to be zero. Hence for $[RuCl_5(NO)]^{2-}$, $2$ p.q.s.(NO$^+$) $= 0.21$ mm/s; for $[Ru(NH_3)_5(NO)]^{3+}$, $0.21$ mm/s $- 2$ p.q.s(NH$_3$) $= +0.36$ mm/s, whence p.q.s.(NH$_3$) $= -0.08$ mm/s.

ligands, since the values for the N-donors are little different from those of the halides. This presumably results from the necessity to use data for organotin compounds, in which the bonding interactions are dominated by the covalency of the Sn—C bonds.

The transition metal carbonyl groups which act as ligands to tin clearly also form very covalent bonds, and function as good donors (McAuliffe, Niven & Parish, 1977$a$, $b$; Dickinson $et$ $al.$, 1975). The donor power of carbonylate anions parallels their nucleophilicities, and increases as carbonyl groups are replaced by phosphines. Detailed examination of data for compounds of the type $SnM_2X_2$ (M = carbonylate anion, X = halide) show considerable discrepancies between calculated and observed values of quadrupole splitting (Dickinson $et$ $al.$, 1975). The calculated values are between 10 and 30% lower than those observed. The trends in the

Fig. 2.14. Partial quadrupole splitting values for ($a$) ruthenium(II), ($b$) iridium(III), and ($c$) gold(I), compared with those for iron(II). The best-fit lines are: p.q.s. $(L)_{Ru} = 0.26$ p.q.s. $(L)_{Fe} + 0.11$ mm/s, $(r^2 = 0.63)$; p.q.s. $(L)_{Ir} = 3.04$ p.q.s. $(L)_{Fe} + 0.81$ mm/s, $(r^2 = 0.86)$; p.q.s. $(L)_{Au} = 2.11$ p.q.s. $(L)_{Fe} - 0.98$ mm/s, $(r^2 = 0.84)$.

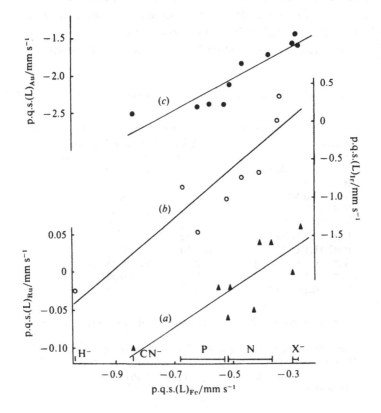

discrepancies is consistent with a change in hybridisation of the tin atom induced by the change in the halogen. As the halogen becomes less electronegative, the $s$ character of the hybrid orbital increases, presumably at the expense of that in the Sn—M bond. The latter bond therefore gains in $p$ character, and p.q.s.$(M)_{Sn}$ should become more negative. The resulting change in electron density at the metal atom is reflected in the C—O stretching frequency. Similar effects have been seen in the $^{119}$Sn and $^{129}$I Mössbauer data for $Me_3SnI$ and $MeSnI_3$, which are paralleled by the NMR coupling constants, $^2J(^{119}SnCH)$ (Marks, Drago, Herber & Potasek, 1976; Jones & Dombsky, 1981).

Such a phenomenon is of course expected. It is unrealistic to imagine that the amount of charge donated by any given ligand would be independent of the identity of the other ligands and of the amount of charge donated by them. It is therefore surprising that the point charge model has enjoyed such considerable success, and there seem to be two possible reasons for this.

Table 2.27. *Partial quadrupole splitting scales for ligands coordinated to tin(IV) in four-, five-, and six-coordination*

| L | p.q.s.$(L)_{Sn}^{tet}$/mm s$^{-1}$ | p.q.s.$(L)_{Sn}^{oct}$/mm s$^{-1}$ | p.q.s.$(L)_{Sn}^{tba}$/mm s$^{-1}$ |
|---|---|---|---|
| $NCS^-$ | $+0.21$ | $+0.07$ | $+0.065$ |
| $Cl^-, F^-$ | $0$ | $0$ | $0$ |
| $Br^-$ | $-0.07$ | $0.00$ | $0.00$ |
| $I^-$ | $-0.17$ | $-0.14$ | $-0.08$ |
| $\frac{1}{2}$ acac$^-$ | | $-0.05$ | $-0.03$ |
| $CH_3CO_2^-$ | $-0.15$ | | $+0.075^a$ |
| py | | $-0.10$ | $-0.03$ |
| $\frac{1}{2}$ bipy | | $-0.08$ | |
| $\frac{1}{2}$ phen | | $-0.04$ | |
| $Co(CO)_4^-$ | $-0.76$ | | |
| $Co(CO)_3(PPh_3)^-$ | $-0.83$ | | |
| $Mn(CO)_5^-$ | $-0.97$ | $-0.71$ | |
| $Mn(CO)_4(PPh_3)^-$ | $-1.01$ | $-0.73$ | |
| $Cr(CO)_3cp^-$ | $-0.72$ | | |
| $Mo(CO)_3cp^-$ | $-0.82$ | $-0.66$ | |
| $Fe(CO)_2cp^-$ | $-1.08$ | $-0.74$ | |
| $H^-$ | $-1.06$ | | |
| $CF_3^-$ | $-0.68$ | | |
| $C_6F_5^-$ | $-0.72$ | | |
| $C_6Cl_5^-$ | $-0.82$ | | |
| $C_6H_5^-$ | $-1.26$ | $-0.95$ | $-0.89$ |
| $CH_3^-$ | $-1.37$ | $-1.03$ | $-0.94$ |

From Parish (1984$a$).
tet = tetrahedral, oct = octahedral, tba = trigonal–bipyramidal axial.
$^a$As bridging ligand.

First, in transition metal systems the bonding must have relatively low covalency so that rehybridisation effects are unimportant. That these effects do exist is shown by the extensive data available on *trans* and *cis* influences (Appleton, Clark & Manzer, 1973), but they are relatively small in systems for which Mössbauer data are available. However, as will be seen later, rehybridisation is readily discernible in the Mössbauer data for gold(I), where the metal—ligand bonds have considerable covalent character. Second, for tin the common ligands fall into two major categories: those for which p.q.s.$(L)_{Sn}$ is large and negative, and those for which it is almost zero. Variations in the latter have very little effect on the calculated quadrupole splitting values, while variations in the former have probably been absorbed during the averaging processes used to calculate p.q.s.$(L)_{Sn}$. It is significant that rehybridisation was only discovered when ligands of intermediate bonding characteristics became available.

### 3.3.3    Townes–Dailey treatment

In the simple case where the Mössbauer atom forms only one bond to a single neighbouring atom, as in iodine(I) derivatives, a more quantitative treatment is possible. In a method developed for the interpretation of NQR data, Townes and Dailey (Dailey & Townes, 1955; see also Das & Hahn, 1958) made three assumptions: (i) the EFG arises entirely within the valence shell of the atom concerned; (ii) the bonding involves only the $p$ orbitals (or has a constant $s$ character); (iii) $\pi$-bonding interactions can be ignored. Then

$$V_{zz} = \frac{-e}{4\pi\varepsilon_0} \sum_i n_i \langle 3\cos^2\theta_i - 1 \rangle_p \langle r_i^{-3} \rangle_p$$

$$= \frac{\frac{4}{5}e}{4\pi\varepsilon_0} \langle r^{-3} \rangle_p [-n_z + \tfrac{1}{2}(n_x + n_y)] \tag{2.25}$$

where $n_i$ is the population of the $p_i$ orbital. The quantity in square brackets is often referred to as the $p$ electron (or $p$ orbital) imbalance, and designated $U_p$. A related quantity, also required for the isomer shift calibrations, is the number of holes in the $p$ shell $h_p$, which is given by

$$h_p = h_x + h_y + h_z = 6 - n_x - n_y - n_z \tag{2.26}$$

Following from this it can be shown that

$$U_p = h_z - \tfrac{1}{2}(h_x + h_y) \tag{2.27}$$

$$\eta = -3(n_z - n_x)/(2U_p) \tag{2.28}$$

For a species with a closed shell configuration, such as an iodide ion, $U_p$ and $h_p$ are both zero, and no quadrupole effects are seen. In an $I_2$ molecule, the configuration of each atom approximates to $5s^2 5p^5$, so that $h_p = U_p = 1$.

The value of $eQ_{127}V_{zz}$ is $-48.2$ mm/s ($-2238$ MHz) and $eQ_{129}V_{zz}$ is equal to $-70.33$ mm/s. Atomic beam measurements give $-2293$ MHz for an isolated $^{127}I$ atom which has the $5s^2 5p^5$ configuration exactly. Linear interpolation between these two fixed points gives $h_p$ equal to $-eQ_{127}V_{zz}/$ (2293 MHz) and allows data for other compounds to be converted to $h_p$ values. Thus, for ICl, $eQ_{127}V_{zz}$ is $-3000$ MHz (average value) corresponding to $h_p = U_p = 1.31$, which implies a charge of $+0.31e$ on the iodine atom. In this case an independent check is available from the $^{35}Cl$ NQR datum $eQ_{35}V_{zz} = 82.5$ MHz. Since $eQ_{35}V_{zz}$ (Cl atom) is 109.7 MHz, $h_p$ for the chlorine atom in ICl is 0.75 and the charge on chlorine is $-0.25e$. Such estimates thus appear to be reasonable within a few percent.

As the atom attached to iodine becomes less electronegative, the charge on the iodine atom becomes less, and this is reflected in the quadrupole coupling constant and the calculated ionicity (Table 2.28). The change in effective electronegativity of carbon or tin in the series $CH_n I_{4-n}$ and $Me_n SnI_{4-n}$ is also evident (Table 2.29). The latter series provide some evidence against the point charge model described above.

Table 2.28. *Ionic character deduced from quadrupole coupling constant values for iodine compounds*

|  | ICl | IBr | $I_2$ | HI |
|---|---|---|---|---|
| $eQ_{127}V_{zz}$/MHz | $-3000$ | $-2892$ | $-2232$ | $-1640$ |
| $h_z = U_p$ | 1.31 | 1.26 | 0.97 | 0.72 |
| Charge on I | $+0.31$ | $+0.26$ |  | $-0.28$ |

Mössbauer data from the compilation by Bancroft & Platt (1972).

Table 2.29. *Bond ionicities in carbon– and tin–iodine derivatives*

|  | $CI_4$ | $CHI_3$ | $CH_2I_2$ | $CH_3I$ |
|---|---|---|---|---|
| $eQ_{127}V_{zz}$/MHz | $-2132$ | $-2051$ | $-1920$ | $-1757$ |
| $h_z$ | 0.93 | 0.89 | 0.84 | 0.77 |
| Charge on I | $-0.07$ | $-0.11$ | $-0.16$ | $-0.23$ |

|  | $SnI_4$ | $MeSnI_3$ | $Me_2SnI_2$ | $Me_3SnI$ |
|---|---|---|---|---|
| $eQ_{127}V_{zz}$/MHz | $-1364$ | $-1270$ | $-1038$ | $-740$ |
| $h_z$ | 0.59 | 0.55 | 0.45 | 0.32 |
| Charge on I | $-0.41$ | $-0.45$ | $-0.55$ | $-0.68$ |

Mössbauer data from Bancroft & Platt (1972), Jones & Dombsky (1981) and Marks *et al.* (1976).

When the asymmetry parameter is non-zero, $h_x = h_y$, and there are too many parameters for direct evaluation. It is then usual to set $n_x = 2.0$ ($h_x = 0.0$), so that $h_y$ and $h_z$ can be obtained (Bancroft & Butler, 1974). For example, for *cis* [PtI$_2$(PPh$_3$)$_2$], $eQ_{127}V_{zz} = -1100$ MHz and $\eta = 0.26$, which gives $U_p = 0.48$, $h_y = 0.08$, $h_z = 0.52$, and $h_p = 0.60$. Data of this type allow assessment of the extent of donation by an iodide ion to a transition metal as a function of the other ligands and, amongst other things, provide evidence for the *trans* influence (Table 2.30).

### 3.3.4    Other calibrations

The conversion of quadrupole splitting data into electron configurations for other isotopes, for which the Townes–Dailey approach is inappropriate, involves less direct methods.

The $^{129}$I data described above were obtained with a $^{129}$Te source which decays to $^{129}$I. Usually, the source is cubic ZnTe, which gives a single line. In an ingenious experiment, single line absorbers were used together with TeO$_2$ or Te(NO$_3$)$_2$ sources, which gave $eQ_{129}V_{zz}$ values for iodine atoms in these matrices (Pasternak & Bukshpan, 1967). Comparison with conventional $^{125}$Te measurements on the same compounds showed that $eQ_{125}V_{zz}/eQ_{127}V_{zz}$ was 0.28. A more recent measurement gives a value of 0.48 for $eQ_{125}V_{zz}/eQ_{129}V_{zz}$ (Langouche *et al.*, 1977). The value of $eQ_{125}V_{zz}$ corresponding to $h_p = 1$ therefore lies between $-642$ and $-772$ MHz ($-22.4$ and $-26.9$ mm/s).

The corresponding quantity for $^{119}$Sn can be derived from the data for matrix-isolated tin atoms in argon (Micklitz & Barrett, 1972). A quadrupole

Table 2.30. *The trans influence in platinum(II) complexes*

|  | $eQ_{127}V_{zz}$/MHz | $\eta$ | $h_z$ |
|---|---|---|---|
| *cis* [PtI$_2$(SbPh$_3$)$_2$] | $-1220$ | 0.18 | 0.56 |
| *cis* [PtI$_2$(AsPh$_3$)$_2$] | $-1190$ | 0.24 | 0.56 |
| *cis* [PtI$_2$(PPh$_3$)$_2$] | $-1110$ | 0.26 | 0.52 |
| *cis* [PtI$_2$(PMePh$_2$)$_2$] | $-1050$ | 0.18 | 0.48 |
| *cis* [PtI$_2$(PMe$_2$Ph)$_2$] | $-1040$ | 0.24 | 0.48 |
| *cis* [PtI$_2$(NH$_2$Oct)$_2$] | $-1190$ | 0.23 | 0.56 |
| *cis* [PtI$_2$($\beta$-picoline)$_2$] | $-1070$ | 0.20 | 0.52 |
| *cis* [PtI$_2$(py)$_2$] | $-946$ | 0.12 | 0.42 |
| *cis* [PtI$_2$(NH$_3$)$_2$] | $-955$ | 0.14 | 0.43 |
| *trans* [PtI(H)(PPh$_3$)$_2$] | $-804$ | 0.34 | 0.39 |
| *trans* [PtI(H)(PEt$_3$)$_2$] | $-788$ | 0.20 | 0.36 |

From Dickinson, Parish & Dale (1980).

split spectrum was attributed to pairs of tin atoms in adjacent substitutional sites. Mutual polarisation of the tin atoms would raise the energy of the $5p_z$ orbital (a crystal field effect), giving a configuration of $5s^2 5p_x^1 5p_y^1$, equivalent to $h = 1$. The observed quadrupole splitting (3.5 mm/s) therefore leads to a lower limit of 7.0 mm/s for $eQ_{119} V_{zz}$ ($h_z = 1$). In covalent compounds there is likely to be some contraction of the valence orbitals relative to the matrix isolated atoms (Pasternak & Barrett, 1981), and a working value of 8.0 mm/s has been suggested (Parish, 1982b).

An equivalent value for $^{121}$Sb is best derived from the good linear correlation found for $eQ V_{zz}$ values for isoelectronic, isostructural compounds of tin and antimony (Bancroft, Butler & Libbey, 1972; Bertazzi, Gibb & Greenwood, 1976); giving $eQ_{121} V_{zz}/eQ_{119} V_{zz} = 3.40$ and 3.19. An appropriate average value of $eQ_{121} V_{zz}$ ($h_z = 1$) is therefore $-26$ mm/s.

These values are summarised in Table 2.31, and their use is illustrated below.

## 4      Bonding in transition metal complexes

### 4.1      *Deductions from partial isomer shift and partial quadrupole splitting scales*

While the p.i.s. and p.q.s. parameters yield valuable information about the degree of covalency in the metal–ligand bond, their value is enhanced when they are considered together.

Both parameters are sensitive to the donor–acceptor characteristics of a ligand, but they respond differently. The partial isomer shift represents the contribution to $s$ electron density at the nucleus, which is increased both by $\sigma$-donation, and by $\pi$-withdrawal of $d$ electron density. The partial quadrupole splitting represents charge placed in the $d$ orbitals, and becomes more negative as $\sigma$-donation increases, but more positive with increase in metal to ligand $\pi$-bonding. This means that

$$\text{p.i.s.(L)} \propto \sigma\text{-basicity of L} + \pi\text{-acidity of L} \tag{2.29}$$

$$\text{p.q.s.(L)} \propto \pi\text{-acidity of L} - \sigma\text{-basicity of L} \tag{2.30}$$

If the two parameters are plotted one against the other, the relative bonding characteristics of the ligands can be compared.

Table 2.31. *Values of the quadrupole coupling constant corresponding to* $h_z = 1$

| Isotope | $^{127}$I | $^{129}$I | $^{125}$Te | $^{121}$Sb | $^{119}$Sn |
|---|---|---|---|---|---|
| $eQ V_{zz}/\text{mm s}^{-1}$ | $-49.4$ | $-70.3$ | $-24.5$ | $-26$ | $-8.0$ |

Fig. 2.15. Correlation diagrams (partial quadrupole splitting against partial isomer shift) for (*a*) iridium(III), and (*b*) ruthenium(II).

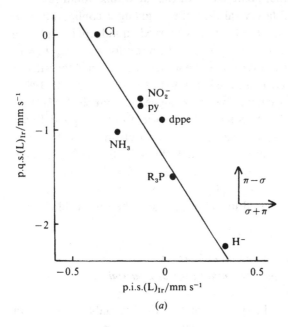

(*a*)

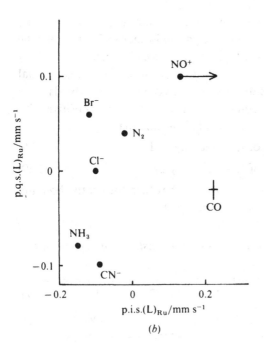

(*b*)

Figure 2.15(*a*) shows such a plot for the limited data available for iridium(III). The parameters show a reasonable negative correlation, suggesting that in this system $\sigma$-effects predominate and increase in the order $L = Cl^- < N$-donor $< P$-donor $< H^-$. Of these, the only ligands which might be thought to have $\pi$-acid properties are the phosphines, but they do not differ significantly from the other ligands. For ruthenium(II) (Figure 2.15(*b*)), there is more scatter, and p.i.s.(L) shows little discrimination, except for CO and $NO^+$, which lie away from the pure $\sigma$-donors in a direction consistent with their $\pi$-acceptor character.

Many data are available for low spin iron(II), and these are shown in Figure 2.16. To simplify the interpretation, the areas corresponding to different types of ligand are shown and, for ease of comparison with the other data, the p.i.s.(L) axis has been reversed. The majority of ligands lie on a broad band of positive slope, suggesting again the predominance of $\sigma$-bonding, with donor power increasing in the order $L = X^- < N$-donor $< P$-

Fig. 2.16. Correlation diagram for low spin iron(II). Note that the partial isomer shift axis is reversed.

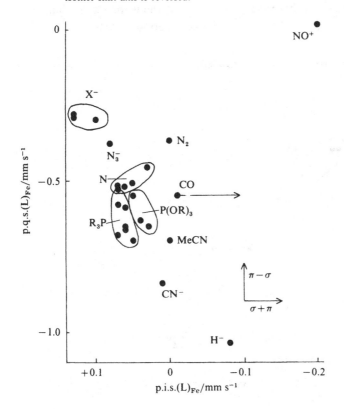

donor$<CN^-<H^-$. The greatest deviation is shown by $NO^+$ which appears as a strong $\pi$-acceptor; synergistic effects may well increase its effective $\sigma$-donor capacity. The carbonyl group is also a good $\pi$-acceptor, and a moderate $\sigma$-donor. On the basis of a single datum, dinitrogen appears to have some $\pi$-acceptor character, but also to be a relatively weak $\sigma$-donor; it may be that $\pi$-donor interactions between the $\pi$-bonding orbitals of $N_2$ and the $t_{2g}$ orbitals of iron partially offset the effects of back donation. Cyanide ion probably functions simply as a strong $\sigma$-donor; $\pi$-bonding effects, as judged here, seem to be neutral. Tertiary phosphines also appear to have little, if any, significant $\pi$-acidity, but both the phosphites $P(OR)_3$ and isonitriles CNR are modest $\pi$-acceptors.

The data for gold(I) (Figure 2.17) give a well-defined band with negative slope, showing the normal order of ligands. For a $d^{10}$ metal ion, $\pi$-bonding effects are not expected, and strong $\pi$-acceptor ligands are usually weakly bound to gold(I). On close inspection, slight curvature is evident, convex upward, which is indicative of a change in hybridisation. When a wider range of data is used, the change in slope is quite evident (Parish, 1982*b*, 1984*a*). This trend implies that the gold atom initially makes relatively greater use of the 6s orbital than the 6p, so that the isomer shift changes faster than the quadrupole splitting. As the ligand donor power is increased,

Fig. 2.17. Correlation diagram for gold(I).

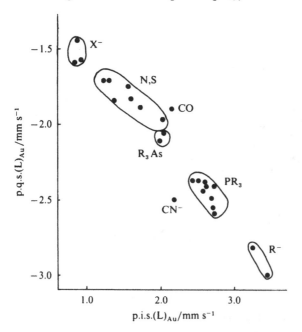

the $6s$ orbital becomes filled and $p$ character increases, so that the rate of increase of quadrupole splitting rises.

### 4.2    *Deductions from ligand parameters*

Data are available for large numbers of transition metal complexes in which the donor atom of the ligand is Mössbauer active: $I^-$, $R_2Te$, $R_3Sb$ and $Cl_3Sn^-$. In every case the parameters of the free ligand are also known, so that the effects of coordination to a particular metal can be assessed, and examined as a function of the oxidation state and the nature of the other ligands (Parish, 1982$a$).

The coordination shifts in the Mössbauer parameters can be converted into numbers of $p$ and $s$ electrons donated to the metal, which are given by

$$\left. \begin{array}{l} P_c = [eQV_{zz}(\text{complex}) - eQV_{zz}(\text{ligand})]/eQV_{zz}(h_z = 1) \\ S_c = [\delta(\text{complex}) - \delta(\text{ligand}) - bP_c]/a \end{array} \right\} \qquad (2.31)$$

where $a$ and $b$ are the isomer shift calibration parameters described in Section 2.6. The total donation of electron density to the metal is then $S_c + P_c$. These quantities are assembled in Figure 2.18, which demonstrates that the same trends in bonding are followed as for conventional

Fig. 2.18. Donation to transition metals by iodide (○), diaryltellurium (●), triarylstibine (▽), and trichlorostannate (▲) ligands. For definitions of the parameters $S_c$ and $P_c$ see above.

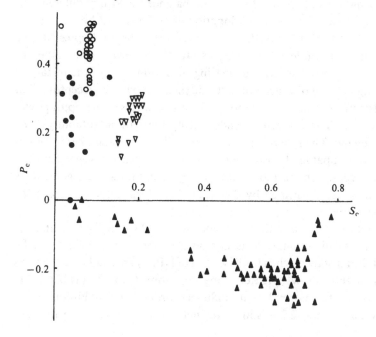

compounds: the iodide and diorganotellurium ligands bond principally through the $5p$ orbital, and there is very little $5s$ participation; total donation is between 0.2 and 0.6 electrons, depending on the acceptor. The organostibines employ both valence orbitals, $5p$ rather more than $5s$, especially with the better acceptors; total donation is between 0.3 and 0.5 electrons. Total donation by $SnCl_3{}^-$ is similar.

For the trichlorostannate ligand data for a wide variety of acceptors are available, and show a pattern quite different from the other ligands. All the $P_c$ values are negative, but $S_c$ is numerically larger and positive. Thus, net donation of electron density occurs in each case, but there is also extensive rehybridisation of the tin atom. In the free ligand, the lone pair has high $5s$ character. On coordination, rehybridisation occurs to increase the $5p$ character, and this becomes more extensive as the acceptor becomes more demanding, i.e. has a higher effective electronegativity (Mays & Sears, 1974). Even after some electron density has been lost to the acceptor (up to 0.5 electrons), the lone pair has preponderant $5p$ character. Eventually a point is reached at which more $5p$ electron density is donated than can be compensated by rehybridisation. With the best acceptors, both $5s$ and $5p$ density are lost, and the slope of the $P_c$ against $S_c$ correlation changes dramatically.

The total charge donated by each of the four marker ligands to a given acceptor is roughly the same, because differences of the substituents and formal oxidation state of the donor atom balance the formal difference in electronegativities. The extent of donation does, however, show a marked dependence on the nature of the acceptor (Parish, 1982c). Representative data are given in Table 2.32 to demonstrate the following generalities: (i) acceptor power increases with increasing oxidation state of the metal; (ii) for these ligands (all of which are 'soft'), $5d$ metals are better acceptors than $4d$ metals; (iii) the presence of carbonyl groups enhances the acceptor power of a metal atom. The last factor is undoubtedly due to the adaptability of the carbonyl ligand. The synergic ($\sigma/\pi$) nature of the bonding makes it a good buffer ligand, capable of absorbing extra electron density when the other ligands are good donors, and compensating when they are less good. This behaviour is also illustrated by the variability of the value of p.i.s.$(CO)_{Fe}$ discussed earlier.

Interpretation of the effects of changes in the ancillary ligands is complicated by directional effects, and both *cis* and *trans* influences can be detected. For instance, when both isomers of $I_2PtL_2$ are available, the *trans* form invariably shows greater donation from I to Pt (Table 2.33), demonstrating the normal *trans* influence order $L = I^- < N\text{-donor} < P\text{-}$donor (see also Table 2.30). Similar deductions, from a narrower range of complexes, have been made from $^{35}Cl$ NQR data (Fryer & Smith, 1970). The high *trans* influences of $H^-$ and $CH_3^-$ are illustrated in Table 2.34

The *cis* influence is often thought to be negligible (Appleton, Clark & Manzer, 1973). However, as shown in Table 2.35, Mössbauer data for the complexes *trans* $I_2PtL_2$ indicate a substantial *cis* influence, increasing in the ligand order, $L = P$-donor $< S$-donor $< N$-donor, which is the reverse of the normal *trans* influence series. This apparent discrepancy with other methods of assessment is probably due to the fact that the latter are mostly based on properties of the whole metal–ligand bond, whereas the Mössbauer data focus attention on the donor atom alone. Calculations of

Table 2.32. *Factors affecting ligand metal donation*[a]

| | $S_c + P_c$ | | $S_c + P_c$ |
|---|---|---|---|
| Change in oxidation state of the metal: | | | |
| $Cl_3SnPtCl(PPh_3)_2$ | 0.34 | $(Cl_3Sn)_2PtCl_2(PPh_3)_2$ | 0.44 |
| Change in principal quantum number of the metal: | | | |
| *cis* $(Ph_3Sb)_2Pd(NO_2)_2$ | 0.45 | *cis* $(Ph_3Sb)_2Pt(NO_2)_2$ | 0.48 |
| $I_2Pd(pdsa)$[b] | 0.47 | $I_2Pt(pdsa)$ | 0.48 |
| $Cl_3SnRh(nbd)_2$[c] | 0.31 | $Cl_3SnIr(nbd)_2$ | 0.35 |
| Replacement of carbonyl groups: | | | |
| $Cl_3SbMn(CO)_5$ | 0.56 | $Cl_3SnMn(CO)_4(PPh_3)$ | 0.42 |
| $(Ph_3Sb)Fe(CO)_4$ | 0.50 | $(Ph_3Sb)_2Fe(CO)_3$ | 0.42 |

[a] The ligand concerned is written first in all formulae.
[b] pdsa $= o\text{-}C_6H_4(SbPh_2)(AsPh_2)$.
[c] nbd $=$ norbornadiene.
From Parish (1982c).

Table 2.33. $S_c + P_c$ *values for* $I_2PtL_2$*: the trans influence*

| L | *trans* $I_2PtL_2$ | *cis* $I_2PtL_2$ | Difference |
|---|---|---|---|
| $PEt_3$ | 0.58 | 0.48 | 0.10 |
| $\beta$-picoline | 0.51 | 0.53 | 0.02 |
| pyridine | 0.50 | 0.46 | 0.04 |
| $NH_3$ | 0.48 | 0.46 | 0.02 |

From Parish (1982c).

the orbital populations on the iodine atom (Armstrong *et al.*, 1976) show good agreement with experimental data (Figure 2.19).

## 5    Summary and conclusions

The information on chemical bonding which can be deduced from the Mössbauer parameters may be summarised as follows.

When the Mössbauer atom is the central atom of the molecule or

Table 2.34. *The trans influence of* $H^-$ *and* $CH_3^-$ *ligands*

|  | $S_c + P_c$ |  | $S_c + P_c$ | Difference |
|---|---|---|---|---|
| *trans* $I_2Pt(PEt_3)_2$ | 0.58 | *trans* $I_2Pt(H)(PEt_3)_2$ | 0.39 | 0.17 |
| *trans* $I_2Pt(PMe_2Ph)_2$ | 0.57 | *trans* $I_2Pt(H)(PMe_2Ph)_2$ | 0.40 | 0.17 |
| *trans* $I_2Pt(PBu_3^n)_2$ | 0.57 | *trans* $I_2Pt(CH_3)(PBu_3^n)_2$ | 0.36 | 0.21 |

From Parish (1982c).

Table 2.35. *The cis influence in trans* $I_2PtL_2$

| L | $PEt_3$ | $PBu_3^n$ | $SMe_2$ | $SEt_2$ | $\beta$-pic | py | $NH_3$ |
|---|---|---|---|---|---|---|---|
| $S_c + P_c$ | 0.58 | 0.57 | 0.55 | 0.51 | 0.51 | 0.50 | 0.48 |

From Parish (1982c).

Fig. 2.19. Calculated and derived $U_p$ values. (After Dickinson, Parish, & Dale, 1980.)

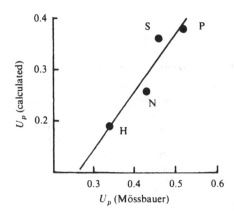

complex:

(a) The extent of donation by its ligands can be monitored, giving the following series:

    (i) for transition metals: $X^- < $ N-donor $<$ P-donor $< CN^- < H^-$, (X = halogen) but the overall variation in covalency is quite small;

    (ii) for gold(I): $X^- <$ N- or S-donor $<$ P-donor $< CN^- < Ar^- < R^-$ (Ar = aryl, R = alkyl), with a large range of covalencies, and some rehybridisation of the gold atom;

    (iii) for tin(IV): $X^- <$ N-donor $\ll Ar^- < R^-$, with a wide range of covalencies and substantial rehybridisation of the tin atom with marked changes in the $s$ participation in the bonding.

(b) For the $p$ block elements, there is increasing employment of the $p$ orbital from tin to iodine.

(c) There is some evidence that covalency of the transition metal–ligand bond increases from $3d$ to $4d$ to $5d$ metals.

(d) Most ligands are predominantly $\sigma$-donors to transition metals, including the tertiary phosphines. The $\pi$-acceptor ability increases from isonitriles and phosphites to carbonyl to nitrosyl ligands.

When the ligand is the Mössbauer probe, trends (b) and (c) are confirmed, as is the substantial rehybridisation involving tin. It is also evident that ligand to metal donation increases with increasing oxidation state and softness of the metal, and when carbon monoxide is also a ligand. Substantial *cis* and *trans* influences can be detected, and these factors can be roughly quantified through the isomer shift and quadrupole splitting calibrations.

It is evident that considerable information about chemical bonding can be derived from the two major Mössbauer parameters, which respond to the amount and distribution of electronic charge on the atom under study. In addition to the obvious factors such as the oxidation state of the Mössbauer atom and the electronegativity of its ligands, the more subtle effects of rehybridisation and, in the case of transition metals, of $\pi$-bonding can be discerned. Reasonable estimates of the charge densities can be made, although independent checks of their validity are not available. Since most of the systems involve heavy metal atoms, reliable molecular orbital calculations are difficult and time consuming. Agreement with other experimental methods is good, especially for the atom-centred methods such as NQR and XPS.

# 3

## *Mössbauer spectroscopy as a structural probe*

GARY J. LONG

### 1    Introduction

The opening chapters of this book have shown how the hyperfine interactions which give rise to the Mössbauer spectrum are a means by which the electronic environment of the nucleus may be examined. The preceding chapter in particular has shown how the Mössbauer parameters can be related to various aspects of chemical bonding and the geometrical properties of compounds. In this chapter the application of Mössbauer spectroscopy in the examination of electronic, molecular, and lattice structure will be considered. Given the close connection between electronic structure and chemical bonding, which has been covered in Chapter 2, this chapter will devote particular attention to molecular and lattice structure, which are both properties amenable to investigation by Mössbauer spectroscopy.

The interpretation of the Mössbauer spectrum obtained from a molecular solid is highly dependent on the particular situation. For example, in some cases two inequivalent Mössbauer atoms may be involved in the molecular complex and this can provide a means by which the Mössbauer data may be more comprehensively interpreted. However, in the majority of compounds studied by Mössbauer spectroscopy only one Mössbauer atom is present and, while acknowledging the advantages of having such a sensitive probe of the immediate local environment, there are often problems in making conclusions concerning the number or disposition of the atoms or groups around the Mössbauer atom. For this reason, Mössbauer studies of a structural nature have been most successful when the Mössbauer data have been complemented by the results from another technique such as X-ray diffraction.

Most of the examples cited in this chapter will be drawn from studies of iron-containing compounds. This is a direct manifestation of the fact that about two-thirds of all Mössbauer studies involve iron (Long, 1984) and the capacity of this element to provide Mössbauer spectra which can be interpreted in terms of molecular and lattice structure. The amenability of the Mössbauer parameters to complement and supplement results obtained from other techniques such as X-ray and neutron diffraction, electron microscopy and solid state nuclear magnetic resonance will also be examined.

## 2        Organometallic compounds
### *2.1    Introduction*

Mössbauer spectroscopy has been a particularly valuable technique for the study of the structural properties of organometallic compounds and useful data has been obtained from compounds containing a variety of Mössbauer atoms. The Mössbauer results have been especially informative when the compounds under examination have contained the Mössbauer atom in more than one site. Several examples of this are provided by studies of iron carbonyls and their derivatives.

### *2.2     $Fe_3(CO)_{12}$*

The history of the structural characterisation of $Fe_3(CO)_{12}$ is an interesting illustration of the successful application of Mössbauer spectroscopy to the determination of the molecular structure of a compound, which had eluded chemists for almost 50 years. Desiderato & Dobson (1982) have written a case history of the determination of the solid state structure of $Fe_3(CO)_{12}$ which is aptly entitled 'The Tortuous Trial Toward the Truth'. The iron carbonyl compound, $Fe_3(CO)_{12}$, was first prepared in 1907 and was soon shown to be diamagnetic, but its molecular formula was not determined until Hieber & Becker (1930) demonstrated that there were three $Fe(CO)_4$ units per molecule. Over the next 30 years a variety of structures, as illustrated in Figure 3.1, were proposed for the molecule, which to varying degrees took into account the experimental evidence available. For instance, the initial infrared spectroscopy studies indicated that $Fe_3(CO)_{12}$ contained two types of carbonyl ligands, which were identified as terminal carbonyl groups, in which the carbon was bonded to a single iron atom, and bridging carbonyl groups, in which either the carbon bridged two iron atoms or the carbon atom was bonded to one iron atom and the oxygen atom was bonded to a second iron atom. Early X-ray structural studies were unsuccessful in determining the structure until Dahl & Rundle (1957) reported preliminary results which eliminated linear structures and indicated a Star of David structure (see Figure 3.1(*g*)). However, because of disorder problems in the molecular structure, the

Fig. 3.1. A variety of proposed structures for $Fe_3(CO)_{12}$.

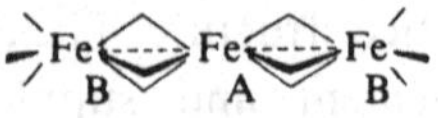

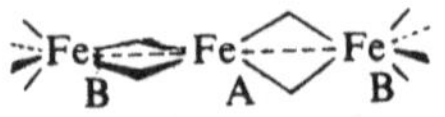

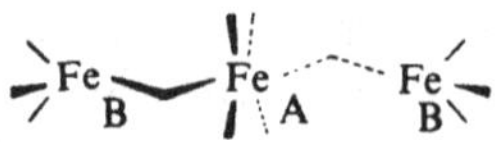

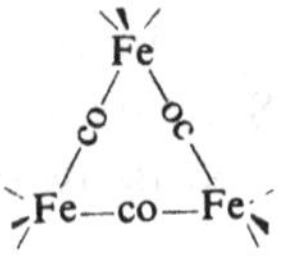

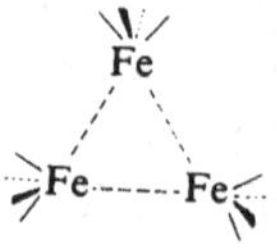

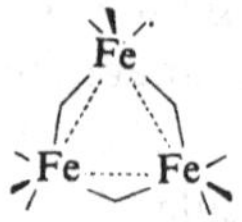

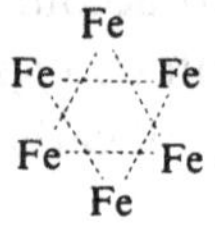

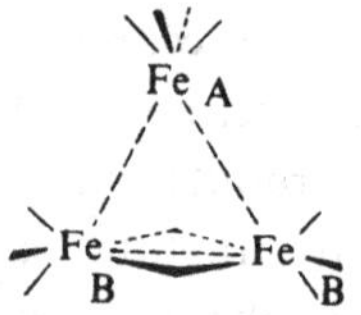

complete single-crystal structure was not revealed until Wei & Dahl (1968) solved the structure.

Early Mössbauer data on $Fe_3(CO)_{12}$ was reported by several investigators (Fluck, Kerler & Neuwirth, 1963; Kalvius, Zahn, Kienle & Eicher, 1962; Herber, Kingston & Wertheim, 1963) and was discussed in detail by Erickson & Fairhall (1965). A summary of some recent results (Long, Russo & Benson, unpublished data; Benson, Long, Kolis & Shriver, 1985) is given in Table 3.1 and illustrated in Figure 3.2. It is immediately apparent from the spectrum obtained at 78 K that there are three absorption lines of roughly equivalent area, perhaps indicating three different isomer shift values for three chemically different iron sites. This possibility may, however, be eliminated immediately because this assignment yields isomer shift values far outside the range typically found for diamagnetic organoiron compounds. A much more reasonable model involves the assignment of the outer lines to a quadrupole split doublet with a relative area of two and the inner line to a doublet with a relative area of one. This is the assignment used in Table 3.1 and shown in Figure 3.2. If this assignment is correct, it must be concluded that $Fe_3(CO)_{12}$ has two different iron sites, one containing only one iron atom, $Fe_A$, in a high symmetry environment, close to octahedral or tetrahedral, which would produce a very small quadrupole interaction, consistent with the observed small splitting of the inner doublet. The second site must have two iron atoms, $Fe_B$, in a much lower symmetry environment, which would produce the larger quadrupole splitting which is observed experimentally.

Table 3.1. *Mössbauer parameters for some organoiron compounds*

| Compound | $T/K$ | Site | $\delta/mm\,s^{-1}$ [a] | $\Delta/mm\,s^{-1}$ | Percentage area |
|---|---|---|---|---|---|
| $Fe_3(CO)_{12}$ | $295^b$ | $Fe_A$ | −0.014 | 0.19 | 45.9 |
| | | $Fe_B$ | 0.022 | 0.94 | 54.1 |
| | $78^c$ | $Fe_A$ | 0.062 | 0.07 | 39.0 |
| | | $Fe_B$ | 0.118 | 1.11 | 61.0 |
| *trans* $[(\eta^5\text{-}C_5H_5)Fe(CO)_2]_2{}^d$ | 296 | — | 0.141 | 1.87 | — |
| | 78 | — | 0.222 | 1.89 | — |
| *cis* $[(\eta^5\text{-}C_5H_5)Fe(CO)_2]_2{}^e$ | 295 | — | 0.123 | 1.92 | — |
| | 78 | — | 0.188 | 1.91 | — |

[a] Relative to iron metal.
[b] From Long, Russo & Benson, unpublished data.
[c] From Benson, Long, Kolis & Shriver (1985).
[d] From Wright, Long, Tharp & Nelson (1986).
[e] From Bryan, Greene, Newlands & Field (1970).

At this point it should be noted that the Mössbauer data alone do not lead to a unique molecular structure for $Fe_3(CO)_{12}$. Although the structures shown in Figure 3.1($d$), ($e$), ($f$), and ($g$) may be eliminated because they have three chemically equivalent sites, the structures given in Figure 3.1($a$), ($b$), ($c$), and ($h$) are all consistent with the Mössbauer spectra. However, by considering the Mössbauer spectra in conjunction with the infrared data, the triangular structure indicated by the preliminary X-ray diffraction results, as well as results for related compounds, such as $Na[Fe_3(CO)_{11}H]$, several of these structures may be eliminated. In this way both Erickson & Fairhall (1965) and Dahl & Blount (1965) were led to the conclusion that the correct structure for $Fe_3(CO)_{12}$ is that shown in Figure 3.1($h$), in which the two $Fe_B$ atoms are bridged by two carbonyl ligands and have three terminal ligands each. This configuration yields a low symmetry electronic environment for $Fe_B$, whereas the single $Fe_A$ site, which has four terminal carbonyl ligands and is bonded to the $Fe_B$ atoms, has a high electronic symmetry.

The correct nature of the proposed structure of $Fe_3(CO)_{12}$ was

Fig. 3.2. The Mössbauer spectra of $Fe_3(CO)_{12}$.

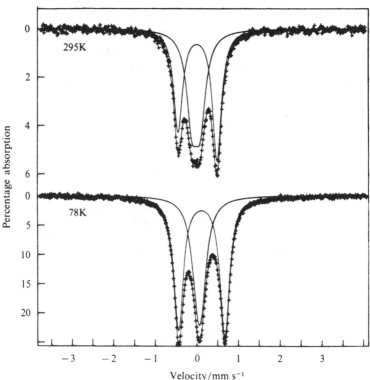

Velocity/mm s$^{-1}$

subsequently confirmed by single-crystal work which agreed with the earlier partial structural analysis which had indicated the Star of David structure shown in Figure 3.1(g). As expected, the structure is disordered, but in an unusual manner in which the three iron atoms are disordered onto the six Star of David sites while the 12 carbonyl ligands remain virtually ordered on their sites. This unusual situation is illustrated in Figures 3.3 and 3.4, which show the two configurations of the iron atoms relative to the fixed framework of the carbonyl ligands.

The study of the structural properties of $Fe_3(CO)_{12}$ provides a good example of the use of Mössbauer spectroscopy for elucidating the structure of a material for which detailed single-crystal X-ray structural results were not available. It also indicates how essential it is for the Mössbauer spectroscopist to consider information derived from other techniques.

It might reasonably be concluded from the above discussion that the solid state structure of $Fe_3(CO)_{12}$ is now well characterised. However, this does not appear to be the case. A close inspection of Figure 3.2 indicates changes in the spectrum with changing temperature. Specifically the quadrupole split doublet shows an increasing area asymmetry, with the inner line also becoming distinctly broadened with increasing temperature. The precise cause of these changes is not well understood but they are likely to be related to the fluctuational behaviour observed in solution infrared and $^{13}C$ nuclear magnetic resonance studies of $Fe_3(CO)_{12}$ and related compounds. Two explanations for this behaviour have been suggested. Cotton & Troup (1974) propose a continuous span of structures, with varying degrees of asymmetric bonding for the two double-bridging carbonyl groups, ranging from a non-bridged configuration (Figure 3.1(e)) to a symmetrically bridged configuration as found in the solid (Figure 3.1(h)). In contrast, Johnson (1976) has proposed a model in which the carbonyl framework remains fixed while the tri-iron core fluctuates within this framework. The changes reflected in the temperature dependence of the Mössbauer spectrum of solid $Fe_3(CO)_{12}$ may well reflect such dynamic processes. After 20 years this is still a potentially fruitful area for further Mössbauer studies.

## 2.3 *Derivatives of* $Fe_3(CO)_{12}$

In a different context, $Fe_3(CO)_{12}$ has proved to be useful as a parent compound in a study (Benson, Long, Kolis & Shriver, 1985) of the electronic changes associated with its reduction to the structurally related methyne and ketenylidene carbide clusters, which may play an important role as intermediates in the reduction of carbon monoxide at a surface.

From the above discussion, it might appear that Mössbauer

Fig. 3.3. The molecular structure of $Fe_3(CO)_{12}$ showing the two disordered configurations of the iron atoms relative to the fixed framework of the 12 carbonyl ligands. Each iron site has an average occupancy of 0.5. (From Desiderato & Dobson, 1982.)

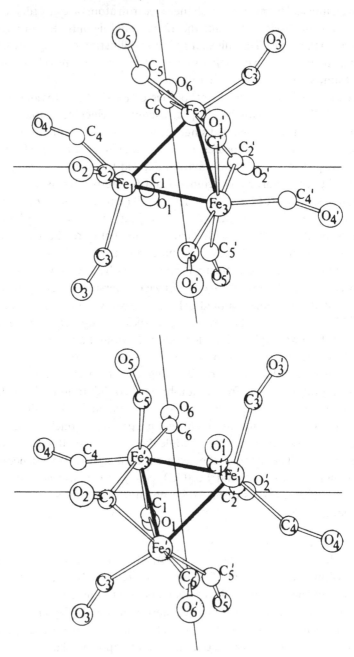

spectroscopy would be quite useful in distinguishing between *cis* and *trans* isomers in organoiron compounds. Although this is sometimes the case (Gibb, Greatrex, Greenwood & Thompson, 1967) it need not always be so, as is illustrated by the results obtained for *cis* and *trans* $[(\eta^5\text{-}C_5H_5)Fe(CO)_2]_2$. The Mössbauer spectrum of the *trans* isomer is shown in Figure 3.5 and the spectral parameters for both compounds are given in Table 3.1 and reveal that the two isomers have virtually indistinguishable parameters. This might well be expected for the isomer shift because both iron atoms in each isomeric dimer have the same coordinated atoms with virtually identical bond distances (Wright, Long, Tharp & Nelson, 1986). This leads to similar $s$ electron density at the iron nucleus and hence similar isomer shifts. It is more difficult to understand why the quadrupole interactions in the two isomers are so similar. Apparently the position of the cyclopentadienyl ring on one iron atom has little influence upon the electric field gradient at the second iron atom in the dimer. This is substantiated by the identical bond angles found in the $Fe_2C_2$ ring in each isomer and the virtually identical iron–iron distance found in each isomer. There is, however, a significant difference in the iron–iron–$C_5H_5$ (centre of gravity)

Fig. 3.4. The superposition of the two iron configurations found within the carbonyl framework in $Fe_3(CO)_{12}$. Each iron site has an average occupancy of 0.5. Six of the terminal carbonyl ligands have been omitted for the sake of clarity. (From Desiderato & Dobson, 1982.)

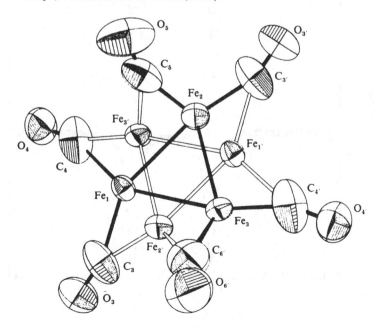

bond angle. The basis for the large quadrupole interaction, as compared with that found in $Fe_3(CO)_{12}$, needs further consideration.

In the above discussion, no mention has been made of the effective atomic number rule proposed by Sidgwick & Bailey (1934). This states that the sum of the number of metallic electrons and those donated by the carbonyls (or other ligands or bonded metals) equals the number of electrons possessed by the next inert gas, in this case the 36 electrons of krypton. This rule, which has proved to be very useful in predicting the electronic and molecular structures of a variety of organometallic compounds, indicates that there should be iron–iron bonds in $Fe_3(CO)_{12}$, shown by the dashed bonds in Figure 3.1, and in both the isomers of $[(\eta^5\text{-}C_5H_5)Fe(CO)_2]_2$. The nature of these metal–metal bonds has been the subject of extensive experimental and theoretical work.

A detailed study of the X-ray and neutron diffraction data for *trans* $[(\eta^5\text{-}C_5H_5Fe(CO)_2]_2$ at 78 K and a subsequent X–N calculation of electronic density deformation maps has yielded an accurate picture of the

Fig. 3.5. The Mössbauer spectra of *trans* $[(\eta^5\text{-}C_5H_5)Fe(CO)_2]_2$.

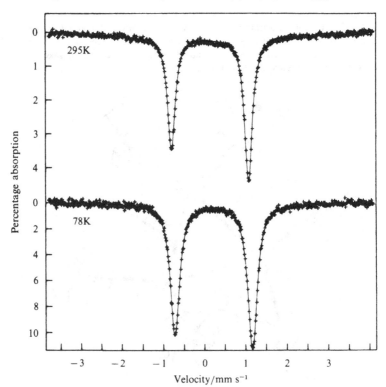

distribution of electronic density throughout this molecule (Mitschler, Rees & Lehmann, 1978). The detailed structure of the molecule is shown in Figure 3.6 and is in excellent agreement with room temperature structural results. In view of the prediction of an iron–iron bond in *trans* [($\eta^5$-$C_5H_5$)Fe(CO)$_2$]$_2$ by the effective atomic number rule, the most surprising result of the electronic density distribution is the lack of any build up of electronic density in the region of the iron–iron bond. A subsequent detailed *ab initio* calculation of the molecular orbital wavefunctions and resulting electronic density distributions have confirmed this finding, which is also supported by similar results for related compounds, which were expected to have, but found not to have, metal–metal bonds (Bénard, 1982). Apparently the direct iron–iron interaction in *trans* [($\eta^5$-$C_5H_5$)Fe(CO$_2$]$_2$ is either non-bonding or somewhat antibonding and the spin coupling between the iron atoms is most likely accomplished through a multicentric molecular orbital made up of the iron *d* orbitals and the $\pi$-antibonding orbitals of the bridging carbonyl ligands. The results of the density distributions are usually presented (Coppens, Li & Zhu, 1983) in terms of the deformation density distribution where

$$\Delta\rho = \rho_{\text{observed}} - \rho_{\text{spherical atom}} \qquad (3.1)$$

and $\rho_{\text{spherical atom}}$ is the sum of the self-consistent field electronic density of

Fig. 3.6. The molecular structure of *trans* [($\eta^5$-$C_5H_5$)Fe(CO)$_2$]$_2$ derived from X-ray and neutron diffraction results obtained at 78 K. (From Mitschler *et al.*, 1978.)

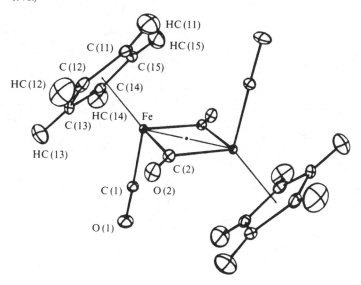

each of the ground state neutral isolated spherically averaged atoms which make up the molecule. The results of these deformation density studies are shown in Figure 3.7.

The *ab initio* self-consistent field electronic population of the iron $3d$ and $4s$ orbitals are presented in Table 3.2, which includes the usual five $3d$ orbitals as well as the spherically averaged six $3d$ orbitals. These results show that the $3d^6 4s^2$ ground state electronic configuration of the neutral iron atom is drastically changed by the bonding in *trans* $[(\eta^5\text{-}C_5H_5)Fe(CO)_2]_2$ to a $3d^{6.98} 4s^{0.066}$ configuration with an electronic charge of $+0.95$. These changes have several important influences on the Mössbauer spectrum of the *trans* isomer. The reduction in the $4s$ orbital occupancy corresponds to a reduction of $4s$ electronic density at the nucleus of the iron atom. In addition the increase in the $3d$ electronic density

Fig. 3.7. (*a*) *Ab initio* self-consistent field calculated electron deformation density distribution of *trans* $[(\eta^5\text{-}C_5H_5)Fe(CO)_2]_2$ in the plane containing the iron–iron axis and the bridging carbonyl ligand. The contour interval is $-200e/nm^3$ and the dashed contours correspond to negative deformation and the solid contours correspond to zero and positive deformation density; (*b*) experimental electron deformation density distribution derived from X–N calculations. The contour interval is $-100e/nm^3$. (From Bénard, 1982.)

(*a*)                        (*b*)

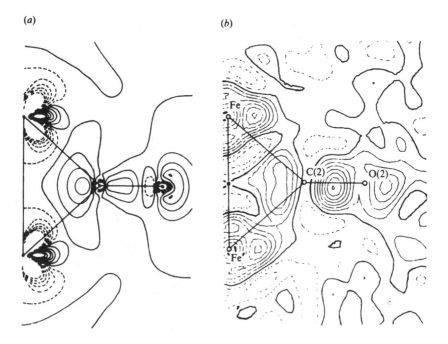

increases the shielding of the $4s$ electrons and thus reduces the effective nuclear charge they experience. This reduction in the effective nuclear charge permits the radial expansion of the $4s$ wavefunction, further decreasing the $s$ electron density at the nucleus (see Chapters 1 and 2 and Shenoy, 1984). Each of these effects are, at least in part, counterbalanced by the increase in the charge on the iron atom. The overall net reduction in the $s$ electron density at the iron nucleus results in the increase in the Mössbauer isomer shift above that found for iron metal (see Table 3.1).

The quadrupole interaction in any material arises from two basic factors, the lattice contribution $(V_{zz})_{lat}$ and the valence contribution $(V_{zz})_{val}$, to the electric field gradient experienced by the $^{57}Fe$ nucleus. The coupling of the nuclear quadrupole moment with the electric field gradient (EFG) partially removes the degeneracy of the $^{57}Fe$ nuclear excited state and leads to the quadrupole split doublet components observed in the Mössbauer spectra of Figures 3.2 and 3.5 (see Chapter 2 for further details). In general, the larger the electric field gradient, the larger will be the quadrupole splitting $\Delta$. It is observed in Table 3.1 that *cis* and *trans* $[(\eta^5\text{-}C_5H_5)Fe(CO)_2]_2$ have substantially larger quadrupole splittings than those observed in $Fe_3(CO)_{12}$. This no doubt results because in the *trans* isomer, and most likely in the *cis* isomer also, there is a substantial contribution from both components of the EFG. In the first case, the lattice contribution results from the different nature and spatial disposition of the three types of ligand. In the second case, the valence contribution results from the

Table 3.2. *Spherically averaged atomic and molecular properties for the iron orbitals in trans* $[(\eta^5\text{-}C_5H_5)Fe(CO)_2]_2$[a]

| Orbital | Spherical atom | Self-consistent field population | Orbital | Spherical atom | Self-consistent field population |
|---------|------------|------------|---------|------------|------------|
| $d_{xy}$ | 1.2 | 1.800 | $d_{xy}$ | 1.2 | 1.800 |
| $d_{xz}$ | 1.2 | 1.912 | $d_{xz}$ | 1.2 | 1.912 |
| $d_{yz}$ | 1.2 | 0.916 | $d_{yz}$ | 1.2 | 0.916 |
| $d_{x^2}$ | 0.8 | 0.372 | $d_{x^2-y^2}$ | 1.2 | 1.095 |
| $d_{y^2}$ | 0.8 | 1.141 | | | |
| $d_{z^2}$ | 0.8 | 0.836 | $d_{2z^2-x^2-y^2}$ | 1.2 | 1.254 |
| $4s$ | 2.0 | 0.066 | $4s$ | 2.0 | 0.066 |

[a] Populations are given in electrons. The $z$ axis is collinear with the iron–iron axis and the $y$ axis is collinear with the bridging carbonyl carbon–carbon axis. Derived from data in Bénard (1982).

inequivalent electron population found in the $3d$ orbitals. If each of these orbitals had the same electronic population, $(V_{zz})_{val}$ would be zero. Such is clearly not the case, as is indicated by the self-consistent field populations of the $d$ orbitals listed in Table 3.2. In fact the population of the $d_{yz}$ orbital is approximately one-half that of the $d_{xy}$ and $d_{xz}$ orbitals. Likewise the density along the $y$ axis (indicated by the $d_{y^2}$ orbital in Table 3.2) is much larger than that along the $z$ axis $(d_{z^2})$ or the $x$ axis $(d_{x^2})$. These differences are clearly apparent in both the theoretical *ab initio* calculation, Figure 3.7($a$), and the experimental plots of the deformation density, Figure 3.7($b$). The low population of the $d_{yz}$ orbital is indicated by the reduced deformation density (broken contour lines) in the vicinity of the iron to bridging carbon vector. The high population along the $y$ axis is indicated by the close solid contours at the immediate right of the iron atoms.

It is difficult to evaluate the relative contributions of $(V_{zz})_{lat}$ and $(V_{zz})_{val}$ to the observed quadrupole interaction in $[(\eta^5\text{-}C_5H_5)Fe(CO)_2]_2$. The similar values of $\Delta$ observed for both the *cis* and *trans* isomers do however indicate similar values for each isomer. In addition the spatial disposition of the ligands on one iron site must have a small influence on the quadrupole interaction on the second iron site. If the influence were large, one would expect to see different quadrupole interactions in the two isomers. This may be an indication that the valence contribution is dominant, a conclusion that would also be supported by the apparent absence of a direct iron–iron bond.

Finally, it should be pointed out that Mössbauer spectroscopy provides further experimental support for the absence of a direct iron–iron bond in $[(\eta^5\text{-}C_5H_5)Fe(CO)_2]_2$. In an unusual $\eta^3$-allyl derivative of *cis* $[(\eta^5\text{-}C_5H_5)Fe(CO)_2]_2$, very different isomer shift values are found for the two different iron sites, values which are different to what would be expected in the presence of a direct iron–iron bond (Cramer, Higa, Pruskin & Gilje, 1983). Similar results have been obtained in some dibridged derivatives of *cis* $[(\eta^5\text{-}C_5H_5)Fe(CO)_2]_2$ (Wright *et al.*, 1986).

## 3    High spin iron(II) coordination complexes

### 3.1    *Introduction*

The several organoiron complexes discussed in the preceding sections are diamagnetic compounds which may be thought of as having essentially covalent bonding involving delocalised molecular orbitals and extensive spin coupling between the metal atoms. It is apparent that Mössbauer spectroscopy is quite sensitive to the coordination number and site geometry of the iron atoms in such compounds. In the following sections we will discuss in detail the properties of the more traditional coordination complexes of iron in which the oxidation state of iron is typically either iron(II) or iron(III), or more occasionally iron(IV) or iron(VI). It will be seen that, for these types of complex, Mössbauer spectroscopy has been found to be sensitive to the nature of the oxidation and spin state, as well as to coordination environment and geometry. In this section the compounds discussed will all have a common iron(II) oxidation state, but a range of coordination environments with different kinds of atoms or ligands coordinated to the iron atom, and with different coordination geometries and spatial arrangements of these ligands. Subsequent sections will deal with other oxidation and spin states.

Before commencing the discussion of specific compounds it will be useful to consider the electronic properties of iron(II) in its different possible environments and how these configurations are reflected in the Mössbauer spectrum of a compound. The iron(II) ion has the $3d^6$ ground state free ion electron configuration. In the presence of a crystal or ligand field the degeneracy of the $^5D$ spectroscopic state arising from the $3d^6$ configuration is removed, producing different results depending upon the geometry of the crystal field. The results, presented both in terms of an orbital splitting diagram and a spectroscopic state diagram, are shown for the octahedral and distorted octahedral case in Figure 3.8 and for the tetrahedral and distorted tetrahedral case in Figure 3.9. These figures are based on the concepts of crystal field theory, which is a conceptually simple approach and is very useful in Mössbauer spectroscopy. The conceptually more complex approach of ligand field theory usually leads to very similar results. Figure 3.8 is valid only for the weak field high spin configuration which results when the crystal field potential is small compared with the interelectronic repulsion between the $3d$ electrons. The strong field low spin configuration and the possible spin crossover case are discussed in Section 4.

In the absence of spin–orbit coupling and in true octahedral or tetrahedral geometry, the first five $3d$ electrons of the partial electronic configuration, $t_{2g}^3 e_g^2$ or $e^2 t_2^3$ respectively, will lead to a spherically symmetric electronic environment and a zero valence contribution to the EFG at the nucleus. The additional sixth electron will have a wavefunction which is a linear combination of the three $t_{2g}$ orbitals, i.e. the $d_{xy}$, $d_{xz}$ and $d_{yz}$ orbitals in octahedral symmetry, or the two $e$ orbitals, i.e. the $d_{x^2-y^2}$ and $d_{z^2}$ orbitals in tetrahedral symmetry. The relative contribution that each of these $d$ orbitals will make to the EFG tensor is presented in Table 3.3. It is apparent from this table that once again the valence contribution to the EFG will be zero. Further, if the lattice contribution to the EFG is also zero, as would be expected in a truly octahedral or tetrahedral lattice site, then the EFG tensor will be zero and the iron(II) ion will exhibit no quadrupole interaction. Hence such a compound will exhibit a single unsplit absorption line in its Mössbauer spectrum.

The situation is quite different, however, if the symmetry at the iron site is reduced below octahedral or tetrahedral symmetry by, for example, the introduction of a tetragonal or trigonal component to the crystal field potential. Such a case might arise if the six octahedral ligands were inequivalent. The result of the lower symmetry is to further reduce the degeneracy of the five $3d$ orbitals as illustrated in Figures 3.8 and 3.9. In this

Fig. 3.8. The crystal field derived orbital splitting diagram for a high spin octahedral iron(II) $3d^6$ complex (*a*), and the corresponding energy level and spectroscopic state diagram (*b*).

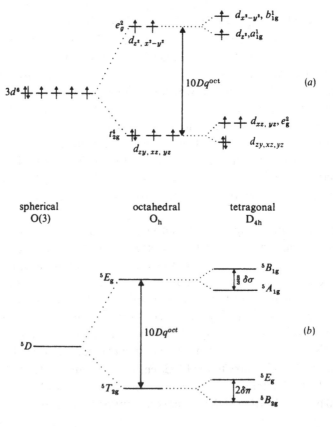

Table 3.3. *The electric field gradient contribution from each 3d orbital*[a]

| Orbital | $V_{zz}$ | $V_{yy}$ | $V_{xx}$ |
|---|---|---|---|
| $d_{x^2-y^2}$ | $-\frac{4}{7}$ | $\frac{2}{7}$ | $\frac{2}{7}$ |
| $d_{z^2}$ | $\frac{4}{7}$ | $-\frac{2}{7}$ | $-\frac{2}{7}$ |
| $d_{yz}$ | $\frac{2}{7}$ | $\frac{2}{7}$ | $-\frac{4}{7}$ |
| $d_{xz}$ | $\frac{2}{7}$ | $-\frac{4}{7}$ | $\frac{2}{7}$ |
| $d_{xy}$ | $-\frac{4}{7}$ | $\frac{2}{7}$ | $\frac{2}{7}$ |

[a] Given in units of $-e\langle r^{-3}\rangle$, where $e$ is the charge on the proton. Adapted from Ingalls (1964).

Fig. 3.9. The crystal field derived orbital splitting diagram for a tetrahedral iron(II) $3d^6$ complex (a), and the corresponding energy level and spectroscopic state diagram (b).

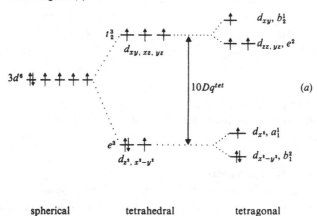

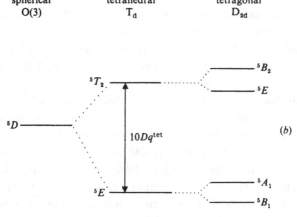

situation, depending upon the magnitude of the low symmetry distortion represented by $\delta\pi$ and $\delta\sigma$ in Figure 3.8 and the temperature $T$, there will be an inequivalent electronic population in the $t_{2g}$ (or $e$) orbitals. The population inequivalence will be given by a Boltzmann distribution in which the inequivalence will be large if $k_B T \ll \delta\pi$ (where $k_B$ is the Boltzmann constant), whereas if $k_B T \gg \delta\pi$ the inequivalence will be very small. The inequivalent population of the three $t_{2g}$ (or two $e$) orbitals will lead to a non-zero value of the EFG tensor as indicated in Table 3.3. It should be noted from this table that the sign of the EFG tensor will depend upon the nature of the orbital ground state.

For the specific case of a tetragonal distortion from octahedral symmetry, the asymmetry parameter $\eta$ of the EFG is zero and the sign of $V_{zz}$ will be positive if the $d_{xy}$ orbital is the lowest in energy (corresponding to a positive value of $\delta\pi$ as shown in Figure 3.8). Alternatively, if $\delta\pi$ is negative, then the $d_{xz}$ and $d_{yz}$ orbitals will be lower in energy than the $d_{xy}$ orbital and the sign of $V_{zz}$ will be negative. Hence, if the sign of $V_{zz}$ can be determined, as for instance through the study of oriented single crystals (Greenwood & Gibb, 1971) or the application of an external applied magnetic field (Reiff, 1973), then it will be possible to determine the ordering of the $3d$ orbitals in a specific complex. A study of the temperature dependence of the quadrupole splitting can be used to evaluate the magnitude of the tetragonal distortion parameter $\delta\pi$ (Dale, 1971). Alternatively a study of a series of related compounds will permit the evaluation of relative values of this parameter for a series of different ligands. It is this last type of study which will be discussed next.

### 3.2     *Iron pyridine complexes*

The various iron(II) complexes of pyridine provide a range of related compounds which illustrate how the Mössbauer parameters respond to changes in coordination environment and coordination geometry as well as structural phase transformations.

The iron(II) halide and pseudohalide coordination complexes formed with pyridine may be divided into two types. The first type consists of complexes with the general stoichiometry, $Fe(pyridine)_4 X_2$, where X represents the halides and pseudohalides listed in Tables 3.4 and 3.5. They are found to be high spin monomeric tetragonally distorted *trans* octahedral complexes and show paramagnetic behaviour down to around 1.3 K (Little & Long, 1978). The typical structure of these complexes is illustrated by that of $Fe(pyridine)_4 Cl_2$ shown in Figure 3.10. The second type, of general stoichiometry $Fe(pyridine)_2 X_2$, can either have a monomeric pseudotetrahedral four-coordinate structure or a polymeric

Table 3.4. *Preparation and structure of iron pyridine complexes*

| | | | Coordination | | |
|---|---|---|---|---|---|
| Reaction | Solvent | X | No. | Environ-ment[a] | Approximate geometry |
| $FeX_2 \cdot xH_2O + 4$ pyridine $\rightarrow Fe(pyridine)_4X_2$[b] | pyridine | Cl, Br, I | 6 | $N_4X_2$ | octahedral monomer[c] |
| $FeSO_4 \cdot 7H_2O + 2NaNCX + 4$ pyridine $\rightarrow Fe(pyridine)_4(NCX)_2$[d] | pyridine | O, S, Se | 6 | $N_4N_2'$ | octahedral monomer[c] |
| $FeX_2 \cdot xH_2O + 2$ pyridine $\rightarrow Fe(pyridine)_2X_2$[e] | methanol | Cl, Br | 6 | $N_2X_4$ | octahedral chain polymer[f] |
| $Fe(pyridine)_4X_2 \rightarrow Fe(pyridine)_2X_2$[e] | none | Cl, Br | 6 | $N_2X_4$ | octahedral chain polymer |
| $Fe(pyridine)_4I_2 \rightarrow Fe(pyridine)_2I_2$[g] | none | I | 4 | $N_2I_2$ | tetrahedral monomer |
| $Fe(pyridine)_4(NCX)_2 \rightarrow Fe(pyridine)_2(NCX)_2$[d] | none | O, S, Se | 6 | $N_2N_2'X_2$[h] | octahedral chain polymer |

[a] N represents the nitrogen of the pyridine ligand and N′ represents the nitrogen of the pseudohalide.
[b] From Little & Long (1978), Long, Whitney & Kennedy (1971) and Long & Clarke (1978).
[c] See Figure 3.10.
[d] From Little & Long (1978) and Reiff, Frankel, Little & Long (1974).
[e] From Little & Long (1978) and Long et al. (1971).
[f] See Figure 3.11.
[g] From Little & Long (1978).
[h] $N_2N_4'$ for $Fe(pyridine)_2(NCO)_2$ in which the NCO bridges only through the nitrogen atom (Little & Long, 1978).

Table 3.5. *Mössbauer parameters for a series of bis and tetrakis iron pyridine complexes*[a]

| Compound | T/K | $\delta$/mm s$^{-1}$ | $\Delta$/mm s$^{-1}$ [b] | Compound | T/K | $\delta$/mm s$^{-1}$ [b] | $\Delta$/mm s$^{-1}$ |
|---|---|---|---|---|---|---|---|
| Fe(pyridine)$_4$Cl$_2$ | 297 | 1.06 | 3.08 | Fe(pyridine)$_2$Cl$_2$ | 297 | 1.08 | 0.57 |
| | 78 | 1.18 | 3.49 | | 233 | 1.11 | 0.55 |
| | 4.2 | 1.10 | 3.42 | | 195 | 1.13 | 0.56 |
| | 1.32 | 1.05 | 3.48 | | | 1.15 | 1.14 |
| | | | | | 78 | 1.21 | 1.25 |
| Fe(pyridine)$_4$Br$_2$ | 297 | 1.03 | 2.20 | Fe(pyridine)$_2$Br$_2$ | 297 | 1.03 | 0.82 |
| | 78 | 1.09 | 2.86 | | 78 | 1.15 | 1.18 |
| | 4.2 | 1.09 | 2.86 | | 4.2 | 1.14 | 1.53 |
| | | | | | 1.10 | 1.15 | 1.54 |
| Fe(pyridine)$_4$I$_2$ | 297 | 0.99 | 0.33 | Fe(pyridine)$_2$I$_2$ | 297 | 0.76 | 0.94 |
| | 78 | 1.11 | 0.53 | | 78 | 0.86 | 1.33 |
| | 4.2 | 1.12 | 0.65 | | 4.2 | 0.95 | 1.81 |
| Fe(pyridine)$_4$(NCO)$_2$ | 297 | 1.04 | 2.43 | Fe(pyridine)$_2$(NCO)$_2$ | 297 | 1.09 | 1.54 |
| | 78 | 1.16 | 2.62 | | 78 | 1.21 | 2.16 |
| | 4.2 | 1.12 | 2.75 | | | | |
| Fe(pyridine)$_4$(NCS)$_2$ | 297 | 1.05 | 1.54 | Fe(pyridine)$_2$(NCS)$_2$ | 297 | 1.02 | 2.60 |
| | 78 | 1.17 | 2.01 | | 78 | 1.12 | 3.02 |
| | 4.2 | 1.16 | 1.90 | | | | |
| Fe(pyridine)$_4$(NCSe)$_2$ | 297 | 1.06 | 0.76 | Fe(pyridine)$_2$(NCSe)$_2$ | 297 | 1.00 | 2.83 |
| | 78 | 1.17 | 0.91 | | 78 | 1.13 | 3.14 |

[a] From Little & Long (1978).
[b] Relative to iron metal.

pseudo-octahedral linear chain structure with two *trans* pyridine ligands and bridging X ligands. Fe(pyridine)$_2$I$_2$ is found to have the pseudotetrahedral structure whereas the other complexes of this type adopt the polymeric linear chain structure, as illustrated in Figure 3.11 for Fe(pyridine)$_2$Cl$_2$.

Selected Mössbauer parameters for the iron pyridine complexes are

Fig. 3.10. A perspective structure of Fe(pyridine)$_4$Cl$_2$, illustrating the tetragonal axis which is coincident with the Fe–Cl direction.

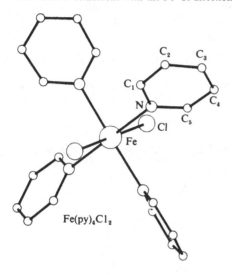

Fe(py)$_4$Cl$_2$

Fig. 3.11. The polymeric structure of Fe(pyridine)$_2$Cl$_2$ observed at room temperature (*a*) and the low-temperature reversible phase transformation to the low-temperature structure (*b*) observed below about 225 K.

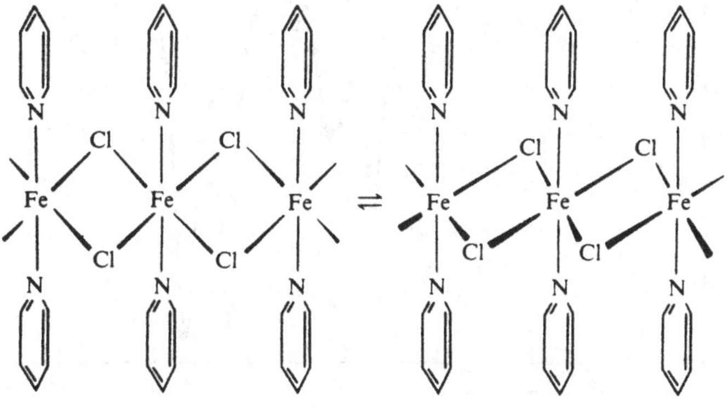

(*a*)                                        (*b*)

given in Table 3.5 and some typical spectra are illustrated in Figures 3.12 and 3.13. In general, all the Fe(pyridine)$_4$X$_2$ and Fe(pyridine)$_4$(NCX)$_2$ compounds exhibit simple two-line quadrupole split doublet spectra from room temperature down to 4.2 K. Values of the isomer shift $\delta$ fall in the range of 0.99 to 1.06 mm/s and increase somewhat with decreasing temperature. This range of values is very typical of six-coordinate octahedral high spin iron(II) complexes with the N$_4$X$_2$ or N$_4$N$_2'$ coordination environment. The increase in $\delta$ with decreasing temperature results from the second-order Doppler shift. In contrast to the isomer shift, the quadrupole splitting $\Delta$ shows a broad range of values between 0.33 and 3.08 mm/s at room temperature and between 0.65 and 3.42 mm/s at 4.2 K. Both the temperature dependence and range of values for $\Delta$ may be

Fig. 3.12. The Mössbauer spectra of polycrystalline Fe(pyridine)$_4$Cl$_2$ at 1.32 K: (a) in zero field and (b) in a transverse applied magnetic field of 0.4 T.

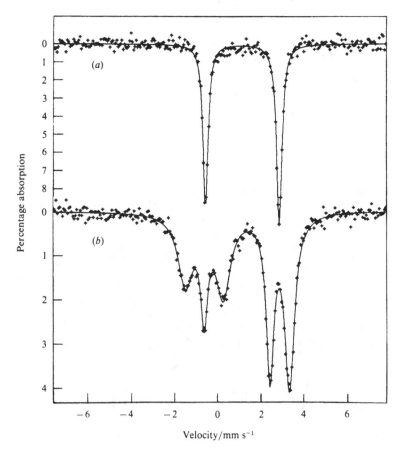

accounted for in terms of the tetragonal distortion (Figure 3.8) of each of these compounds (Little & Long, 1978). Apparently the bonding characteristics of the chloride ligand when compared with the pyridine ligand yields a much larger tetragonal component to the crystal field potential than does the iodide ligand. Because the $t_{2g}$ orbitals in an octahedral complex bond in a $\pi$-bonding fashion (Purcell & Kotz, 1977) it is possible to relate the extent of their splitting by the tetragonal distortion (see Figure 3.8(*a*)) in terms of the angular overlap $\pi$-bonding factors, $\varepsilon_{\pi py}$. A

Fig. 3.13. The Mössbauer spectra of polycrystalline Fe(pyridine)$_2$Cl$_2$.

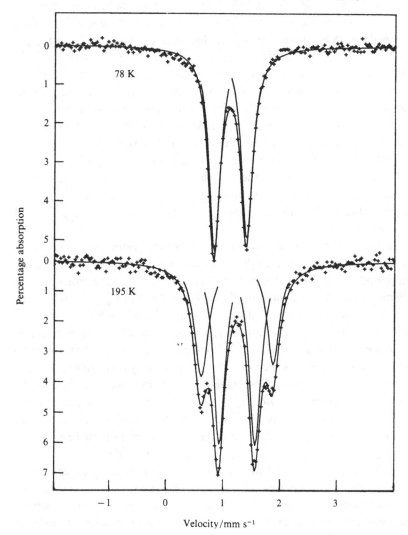

Velocity/mm s$^{-1}$

combination of X-ray structural studies (Long & Clarke, 1978) and applied magnetic field Mössbauer spectra have indicated that all the tetrakis pyridine complexes are tetragonal and have a positive value for $V_{zz}$, the principal component of the EFG (Little & Long, 1978). This is clearly seen in Figure 3.12(b) which shows that an applied field splits the high velocity line in $Fe(pyridine)_4Cl_2$ into a doublet and the low velocity line into a triplet. The positive value of $V_{zz}$, as shown in Table 3.3, corresponds to a $^5B_{2g}$ electronic ground state as shown in Figure 3.8(b). If it is assumed that $\varepsilon_{\pi py}$ is constant in each of the tetrakis pyridine complexes, an assumption that is supported by the constancy of the iron–nitrogen bond distance in these complexes, the relative values of $\varepsilon_{\pi L}$ may be determined from the observed quadrupole splittings. By using the quadrupole splitting values obtained at 78 K, it may be concluded that $\varepsilon_{\pi Cl} > \varepsilon_{\pi Br} > \varepsilon_{\pi NCO} > \varepsilon_{\pi NCS} > \varepsilon_{\pi NCSe} > \varepsilon_{\pi I} \sim \varepsilon_{\pi py}$. The chloride ligand is most effective in stabilising the $d_{xy}$ orbital (the $^5B_{2g}$ state) relative to the $d_{xz}$ and $d_{yz}$ orbitals (the $^5E_g$ state). Thus the $\pi$-bonding is most effective between the chloride orbitals and the iron orbitals and is least effective for the pyridine and iodine orbitals. A study (Little & Long, 1978) of the electronic spectral properties of these compounds has permitted a similar evaluation of the relative $\sigma$-bonding properties of these ligands yielding $\varepsilon_{\sigma py} > \varepsilon_{\sigma NCSe} > \varepsilon_{\sigma NCS} > \varepsilon_{\sigma Cl} > \varepsilon_{\sigma NCO} > \varepsilon_{\sigma Br} > \varepsilon_{\sigma I}$. Thus a study of both the Mössbauer spectra and the electronic absorption spectra of a homologous series of compounds has allowed the determination of the relative $\pi$- and $\sigma$-bonding characteristics of a series of different ligands.

The Mössbauer spectral parameters for the bis pyridine complexes are of interest because they reveal differences in the structural properties of the compounds. The room temperature isomer shifts range from 1.00 to 1.09 mm/s for all these compounds, except $Fe(pyridine)_2I_2$, for which the value of $\delta$ is only 0.76 mm/s. This much smaller isomer shift value is clearly indicative of the reduced coordination number found in this and related pseudotetrahedral compounds (e.g. Long & Coffen, 1974). The tetrahedral coordination geometry, which is further supported by X-ray diffraction and electronic spectral studies, presumably results from the large size of the iodide ions as compared with the other halide and pseudohalide ligands.

The quadrupole splitting observed in the bis pyridine complexes also indicates the nature of the coordination environment in these linear chain bridging ligand complexes. This is observed in terms of the large $\Delta$ values found for $Fe(pyridine)_2(NCS)_2$ and $Fe(pyridine)_2(NCSe)_2$ as compared with the other bis complexes. These large values arise because in these two compounds the coordinated atoms are two *trans* pyridine nitrogen atoms, two pseudohalide nitrogen atoms, and two sulphur or selenium atoms from the bridging NCS or NCSe, giving a coordination environment of $N_2N_2S_2$ or $N_2N_2Se_2$, a decidedly lower symmetry environment than the $N_2X_4$ environment found in $Fe(pyridine)_2Cl_2$ and $Fe(pyridine)_2Br_2$. The $Fe(pyridine)_2(NCO)_2$ compound has an intermediate splitting indicative of the bridging of the NCO ligand solely through the nitrogen atom.

The Mössbauer spectra of $Fe(pyridine)_2Cl_2$, shown in Figure 3.13, illustrates a large change in the quadrupole splitting between 195 and 78 K and the existence of two different iron sites at 195 K. There is, however, little change in the isomer shift. The large change in $\Delta$ and the observation of two doublets at intermediate temperatures are indicative of the occurrence of a structural phase transition in this compound. This first-order crystallographic phase transition, which corresponds to a change from symmetric to asymmetric bridging chloride ligands, as shown in Figure 3.11, has also been observed by single-crystal X-ray diffraction (Clarke & Milledge, 1975) and magnetic anisotropy (Bentley, Gerloch, Lewis & Quested, 1971) studies of $Co(pyridine)_2Cl_2$. An applied field Mössbauer study of $Fe(pyridine)_2Cl_2$ has shown that $V_{zz}$ is positive in both the $\alpha$ and $\gamma$ phases, indicating that the change in $\Delta$ results from the change in symmetry at the iron site rather than a change in the electronic ground state of the complex. If the compound is prepared by precipitation in methanol the two quadrupole split doublets are observed to coexist between 217 and 235 K. However, when obtained by thermolysis, the phase transition does not go to completion, presumably as a result of the defective nature of the thermolytic product, and both doublets are observed at 78 K. Related Mössbauer studies of $^{57}Fe$-doped $Mn(pyridine)_2Cl_2$ and $Ni(pyridine)_2Cl_2$ indicate a similar structural change (Yoshihashi & Sano, 1975). Further evidence for the first-order nature of this phase transition has been obtained from differential scanning calorimetry which indicated that the transition has an endothermic latent heat of between 5 and 19 J/mol. The transition was found to occur at 254, 250 and 164 K in $Mn(pyridine)_2Cl_2$, $Fe(pyridine)_2Cl_2$ and $Co(pyridine)_2Cl_2$ respectively (Eismann & Reiff, 1980).

An interesting application of Mössbauer emission spectroscopy is illustrated by the study of $^{57}Co(pyridine)_2Cl_2$ which was used as a source against a single-line absorber. The resulting emission spectra clearly reveal the presence of the phase transition and indicate that this polymeric material is able to withstand the damage associated with the electron capture decay of the $^{57}Co$.

## 3.3 *Phthalocyanine iron(II)*

The phthalocyanine iron(II) complex, Fe(pc), illustrated in Figure

3.14, provides a good example of iron(II) in a square-planar coordination enrivonment and Mössbauer spectroscopy has proved to be very useful in elucidating its electronic properties. These are of considerable interest as they relate to the understanding of the structure and function of iron haem proteins.

There has until recently been considerable controversy about the ground state electronic environment produced by the phthalocyanine ligand. As the symmetry is reduced from octahedral to square planar the remaining degeneracy of the octahedral $t_{2g}$ and $e_g$ orbitals is removed. The nature of the final ordering of these states will then depend upon the relative $\pi$- and $\sigma$-bonding characteristics of the ligand. Three of the possible electronic configurations for the $3d$ orbitals are presented in Figure 3.15. Each of these configurations is consistent with the early magnetic work (Klemm & Klemm, 1935; Lever, 1966) on Fe(pc) which indicated an intermediate $S = 1$ spin state with two unpaired electrons. It is therefore necessary to determine which of the possibilities shown in Figure 3.15 is correct. Unfortunately, studies involving different techniques appear to provide different asnwers, with nuclear magnetic resonance and magnetic studies indicating the $^3A_g$ ground state favoured by self-consistent field molecular orbital calculations, while low-temperature magnetisation and paramagnetic anisotropy measurements point to the $^3B_{2g}$ ground state (Dedieu, Rohmer & Veillard, 1982).

Mössbauer studies of Fe(pc) in zero applied magnetic field have been carried out by Hudson & Whitfield (1967) and Moss & Robinson (1968).

Fig. 3.14. The molecular structure of phthalocyanine iron(II), Fe(pc). (From Kirner *et al.*, 1976.)

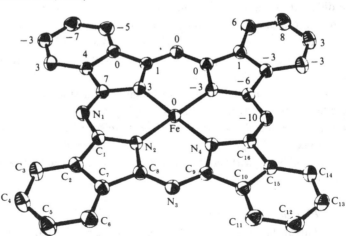

The 78 K Mössbauer spectrum of Fe(pc) is shown in Figure 3.16. Unfortunately the Mössbauer spectra do not help in resolving the conflict over the ground state configurations. However, an applied field Mössbauer study as well as low-temperature magnetic work by Dale and coworkers

Fig. 3.15. Three possible electronic configurations for the Fe(pc) molecule.

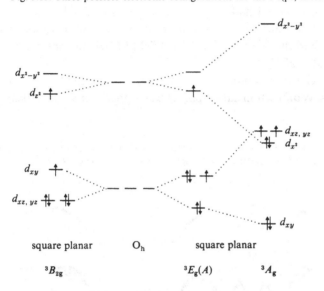

| square planar | $O_h$ | square planar |
|:---:|:---:|:---:|
| $^3B_{2g}$ | $^3E_g(A)$ | $^3A_g$ |

Fig. 3.16. The zero field Mössbauer spectrum of Fe(pc) obtained at 78 K.

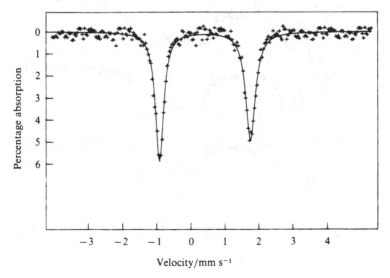

(Dale, 1974) has indicated the $^3E_g(A)$ ground state. The applied field Mössbauer spectra illustrated in Figure 3.17 are consistent only with a positive value of $V_{zz}$ and a small value of the asymmetry parameter. Similar results were obtained from single-crystal experiments, but there was also found to be a significant departure from tetragonal symmetry, which will further remove the degeneracy of the $d_{xy}$ and $d_{yz}$ orbitals shown in Figure 3.15. By carefully considering the contributions that each of the possible configurations would make to the EFG it was concluded that only the $^3E_g(A)$ ground state is consistent with both the low-temperature magnetic properties and the fact that the applied field Mössbauer spectra require a positive value of $V_{zz}$.

This conclusion is supported by some recent accurate low-temperature (110 K) X-ray diffraction data reported by Coppens, Li & Zhu (1983) which

Fig. 3.17. The Mössbauer spectrum of Fe(pc) obtained at various temperatures in a 3.0 T transverse applied magnetic field. (From Dale, 1974.)

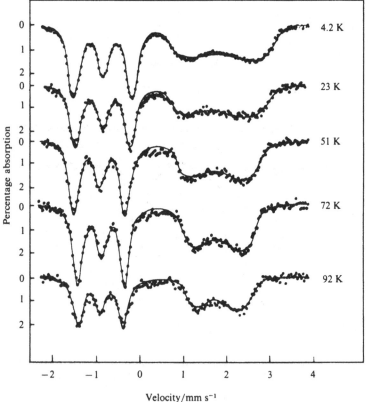

indicate that the low-temperature structure is essentially the same as that obtained at room temperature (Kirner, Dow & Scheidt, 1976). The electron deformation density map indicates that the excess bonding density over the isolated spherical atom is located along the bisectors of the iron–nitrogen bonds, which eliminates the $^3B_{2g}$ configuration. More useful are the experimental multipole populations (Holladay, Leung & Coppens, 1983) which yield the $d$ orbital occupancies given in Table 3.6, which also contains the predicted occupancies for each of the configurations shown in Figure 3.15. These results show the occupancy ratios of the $d_{xy}$, $d_{xz, yz}$ and $d_{z^2}$ orbitals are 2:2.6:1.1 respectively, which is only compatible with the $^3E_g(A)$ state as suggested by Dale (1974) for the ground state electronic configuration of Fe(pc). The experimentally observed non-zero occupancy of the $d_{x^2-y^2}$ orbital is evidence for covalency in the iron to phthalocyanine bonding.

### 3.4 Phase transitions in iron(II) hydrates and ammoniates

The Mössbauer spectral properties of the various hydrate and ammoniate complexes that can be formed with iron(II) are interesting because of the information they provide about the phase transformations or order–disorder transformations which these compounds undergo. Mössbauer data and details of the phase transitions are given in Table 3.7.

In an extension of the early work of Dezsi, Keszthelyi, Molnar & Pocs (1965 and 1968), Brunot has studied the temperature dependence of the isomer shift and quadrupole interaction in a series of related hydrates and deuterates. All of the compounds show quadrupole split doublet Mössbauer spectra and, as indicated in Table 3.7 and illustrated in Figure 3.18, the isomer shift proves to be a sensitive indication of the extent of hydration of iron(II) chloride. This serves as a powerful illustration of the sensitivity of this parameter to the coordination environment and the change in that environment as chloride is gradually displaced by the oxygen of water. With the exception of $FeCl_2 . 2H_2O$, no change in the isomer shift was observed upon deuteration of the water. The small shift in the dihydrate may be related to the linear chain structure of this type of material and the influence of hydrogen bonding between the chains (Narath, 1965). The quadrupole splitting varied smoothly with changing temperature for each of the hydrates, indicating that there are at most only small changes in the site symmetry or more likely a change in the Boltzmann population of the orbitals making up the ground state configuration. The nature of the quadrupole interaction in $FeCl_2 . 4H_2O$ has been discussed in detail by Spiering (1984). Once again, there are, with the exception of $FeCl_2 . 6D_2O$, only small changes in $\Delta$ upon deuteration,

Table 3.6. *Experimental and predicted occupancies for the 3d orbitals of* Fe(pc)[a]

| 3d orbital | Spherical atom occupancy | Experimental occupancy | Prediction for $^3E_g(A)$ | Prediction for $^3A_g$ | Prediction for $^3B_{2g}$ |
|---|---|---|---|---|---|
| $d_{x^2-y^2}$ | 1.2 | 0.75(6), 13.9% | 0 | 0 | 0 |
| $d_{z^2}$ | 1.2 | 0.88(6), 16.3% | 1, 17% | 2, 33% | 1, 17% |
| $d_{xz,yz}$ | 2.4 | 2.13(6), 39.4% | 3, 49% | 2, 33% | 4, 67% |
| $d_{xy}$ | 1.2 | 1.65(8), 30.5% | 2, 33% | 2, 33% | 1, 17% |
| $V_{zz}$ | 0 | >0 | >0 | $\approx 0$ | <0 |

[a] Adapted from Coppens et al. (1983).

Table 3.7. *Mössbauer and other parameters for a series of iron(II) hydrates and ammoniates*[a]

| Compound | Isomer shift $\delta$/mm s$^{-1}$[b] | | | Quadrupole splitting $\Delta$/mm s$^{-1}$ | | | Phase transformation[d] | |
|---|---|---|---|---|---|---|---|---|
| | Room temperature | 100 K | Isotope effect[c] | Room temperature | 100 K | Isotope effect[c] | $T_c$/K | Hysteresis |
| FeCl$_2$.1H$_2$O | 0.96 | 1.07 | no | 2.15 | 2.59 | small | none | — |
| FeCl$_2$.2H$_2$O | 0.99 | 1.09 | yes | 2.28 | 2.61 | yes | none | — |
| FeCl$_2$.4H$_2$O | 1.05 | 1.15 | no | 3.01 | 3.09 | yes | none | — |
| FeCl$_2$.6H$_2$O | 1.03 | 1.13 | no | 1.58 | 1.69 | yes | <100 | |
| FeCl$_2$.6D$_2$O | 1.03 | 1.13 | no | 1.60 | 1.79 | yes | 126 | no |
| Fe(ClO$_4$)$_2$.6H$_2$O | 1.09 | 1.19 | small | 1.34 | 3.48 | no | 252 | ±5 |
| Fe(ClO$_4$)$_2$.6D$_2$O | 1.08 | 1.18 | small | 1.34 | 3.48 | no | 256 | ±5 |
| Fe(BF$_4$)$_2$.6H$_2$O | 1.09 | 1.20 | small | 1.29 | 3.45 | yes | 276 | ±4 |
| Fe(BF$_4$)$_2$.6D$_2$O | 1.10 | 1.20 | small | 1.31 | 3.46 | yes | 283 | ±2 |
| FeCl$_2$.6NH$_3$ | 1.04 | 1.18 | [e] | 0.0 | 1.50 | [e] | 112 | ±4 |
| FeBr$_2$.6NH$_3$ | 1.04 | 1.18 | [e] | 0.0 | 0.0[f] | [e] | 45 | ±2 |
| Fe(ClO$_4$)$_2$.6NH$_3$ | [g] | [g] | [e] | 0.0 | 0.95 | [e] | 157 | ±2 |
| Fe(BF$_4$)$_2$.6NH$_3$ | [g] | [g] | [e] | 0.0 | 1.78 | [e] | 140.5 | ±1.5 |
| FeI$_2$.6NH$_3$ | 1.04 | 1.18 | [e] | 0.0 | 1.25 | [e] | 228 | no |
| Fe(NO$_3$)$_2$.6NH$_3$ | 1.04 | 1.18 | [e] | 0.0 | 1.85 | [e] | 99 | yes |

[a] From Brunot (1974) and Asch *et al.* (1975*a,b*).

[b] Relative to iron metal.

[c] The change from hydration to deuteration.

[d] The average of the heating and cooling transition temperature $T_c$ is given.

[e] Not determined.

[f] A value of 0.7 mm/s is observed at 4.2 K.

[g] Not specified but indicated as similar to FeCl$_2$.6NH$_3$.

which may be understood in terms of the variation in lattice parameters and hydrogen bonding following deuteration. In $FeCl_2 . 6D_2O$ there is a reversible discontinuity in the quadrupole interaction at 126 K, indicating a change in symmetry at the iron site as discussed below. A related change in $FeCl_2 . 6H_2O$ presumably occurs below about 100 K, the lowest temperature studied by Brunot (1974).

The situation is different in the hexa-aquo complexes of iron(II) perchlorate and tetrafluoroborate. Each of these complexes exhibits a sharp increase in the quadrupole interaction at a temperature slightly below room temperature. This change is observed in both the hydrates and deuterates. The dramatic change in the quadrupole interaction is immediately obvious in the Mössbauer spectra of $Fe(BF_4)_2 . 6H_2O$ at 295 and 78 K shown in Figure 3.19, and the temperature dependence shown in Figure 3.20. The change occurs with a small temperature hysteresis and at a slightly higher temperature in the deuterated compound. The transformation appears very similar in the perchlorate salt, but occurs at about 250 K, which is 20

Fig. 3.18. The temperature dependence of the Mössbauer isomer shift for several hydrates of iron(II) chloride. (Adapted from Brunot, 1974.)

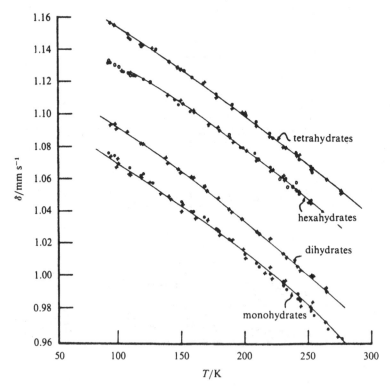

degrees lower than in the tetrafluoroborate salt. These results are all consistent with a first-order phase transformation in which the iron experiences a higher symmetry at higher temperature. The structural changes are almost certainly associated with the order–disorder of the anions. At high temperature the perchlorate and tetrafluoroborate anions are free to rotate or oscillate on their lattice sites, disrupting or at least reducing the hydrogen bonding between the coordinated water and the anion. A similar situation is observed in some related titanium complexes (Figgis & Wadley, 1972). Apparently this anion disorder yields a smaller EFG at the iron site. At lower temperatures the anions are frozen or locked into a fixed position and form hydrogen bonds to the coordinated water. The small differences observed in the transition temperature for the hydrates and deuterates may be accounted for in terms of an isotope effect and the larger difference in transition temperature for the different anions results because of the stronger O—H—F hydrogen bonding with the $BF_4$ anion than the O—H—O interaction with the $ClO_4$ anion. It is interesting

Fig. 3.19. The Mössbauer spectra of $Fe(BF_4)_2 . 6H_2O$.

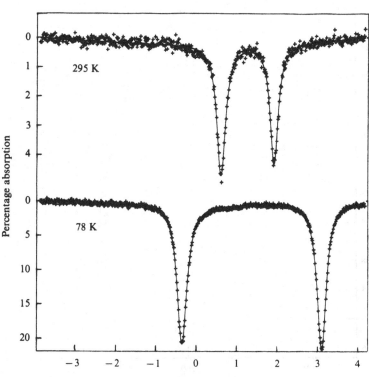

Percentage absorption

Velocity/mm s$^{-1}$

that the quadrupole splitting in the two compounds is so similar both above and below the transition (see Table 3.7). This is probably related to the similar nature of the $[Fe(H_2O)_6]^{2+}$ cations and the fact that the structural order–disorder does not occur directly in the coordinated ligands.

A different situation has been observed (Asch *et al.*, 1975a) in the ammine salts, $FeCl_2 \cdot 6NH_3$ and $FeBr_2 \cdot 6NH_3$, where a change occurs directly on the coordinated ligand, $NH_3$. The Mössbauer spectra show a dramatic change with temperature such that a single absorption line is found at room temperature and a doublet at lower temperatures. Figure

Fig. 3.20. The temperature dependence of the quadrupole splitting in $Fe(BF_4)_2 \cdot 6H_2O$. (Adapted from Brunot, 1974.)

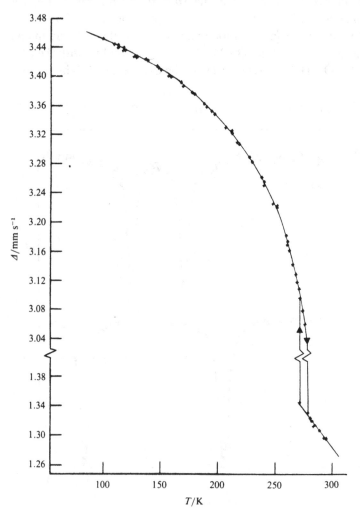

3.21 shows the Mössbauer spectra of $FeCl_2 . 6NH_3$ and it can be seen that there is a single line down to 109.5 K. Between 109.5 and 107 K three lines, a singlet and a quadrupole split doublet, are observed, whereas below 107 K only the doublet is present. The bromide is similar in behaviour except that the change occurs at around 45 K. The electrostatic interactions between the hydrogen atoms in hexa-ammine complexes has been studied by Bates & Stevens (1969). Their work, based on a point charge model, has predicted

Fig. 3.21. The Mössbauer spectra of $FeCl_2 . 6NH_3$. (Adapted from Asch *et al.*, 1975*a*.)

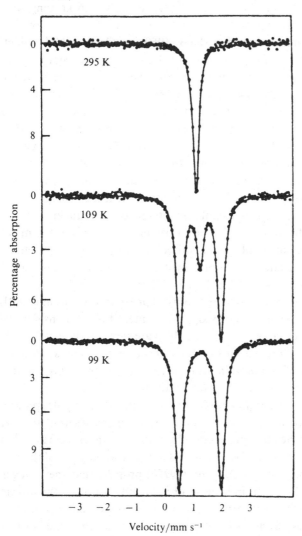

Velocity/mm s$^{-1}$

a phase transition in these materials as well as low-temperature heat capacity anomalies. They show that a cubic site symmetry at the central iron site is not possible if the six ammonia ligands are static. If they were static at least a trigonal component to the site symmetry at the metal would be generated with a resultant EFG. Hence the failure to observe a room temperature quadrupole interaction at the iron site in $FeCl_2 \cdot 6NH_3$, and in several related complexes, indicates that the ammonia ligand must have dynamic properties at higher temperatures (see Table 3.7). This behaviour suggests that the ammonia molecule must be rotating with a frequency above $10^7$ Hz, the minimum frequency which would average the quadrupolar interaction to zero on the Mössbauer timescale. This is in keeping with electron paramagnetic resonance studies (Bates & Stevens, 1969) on $NiCl_2 \cdot 6NH_3$ and infrared studies (Kim, 1960) on $CoCl_2 \cdot 6NH_3$ which suggest that the rotation rate is about $10^{13}$ Hz. At lower temperatures the rotational motion is apparently hindered and the point symmetry at the metal is reduced below that of cubic symmetry. The electrostatic calculations on one cluster indicate several energy minima which may be overcome with thermal energy. However, to account for the critical behaviour at the phase transition, Bates & Stevens (1969) were forced to consider the interaction of the cluster with its 12 near neighbour clusters. The minimal energy occurs when all 13 clusters have the same configuration (Bates, 1970). At low temperature the lattice energy of the system is determined by near neighbour interactions producing a long-range preferred distortion, which yields a trigonal symmetry at the metal site. This dependence of the phase transition upon long-range interactions indicates why such a large range of transition temperatures is observed when the anion is changed (Table 3.7).

In concluding this discussion of the study of phase transitions in iron(II) complexes by Mössbauer spectroscopy, it is interesting to consider the spectra obtained from $Fe(NO_3)_2 \cdot 6NH_3$ which are shown in Figure 3.22. At temperatures above 99 K the Mössbauer spectra are very similar to those found for the related hexa-ammine complexes with other anions, and the ammonia rotational order–disorder phase transformation occurs at 228 K. However, below 99 K a second doublet with a smaller quadrupole splitting appears. The relative intensity of this inner doublet increases until at 4.2 K it comprises 80% of the absorption area. All of these changes were found to be reversible, but a hysteresis was observed in the relative areas of the two doublets with temperature. Asch *et al.* (1975*b*) propose that the changes observed below 99 K are an indication of hindered motion of the nitrate ions. At high temperatures the triangular nitrate groups oscillate around an oxygen–oxygen edge, while at lower temperatures the hindrance of this oscillation fixes the nitrate ion in one or the other of two positions and yields a static distortion of two types. As the temperature is further lowered one of the two sites becomes preferred and the population of the two sites changes slowly with decreasing temperature. This accounts for the two

Fig. 3.22. The Mössbauer spectra of $Fe(NO_3)_2 . 6NH_3$. (Adapted from Asch *et al.*, 1975*b*.)

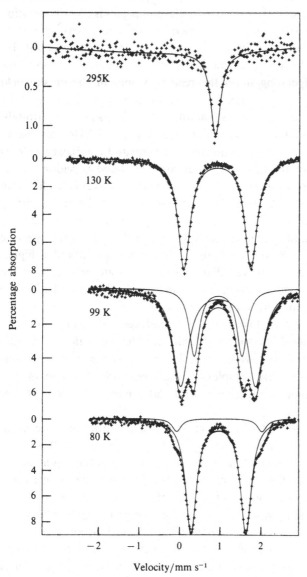

Velocity/mm s$^{-1}$

quadrupole split doublets which are observed to change their relative intensity on cooling, and for the hysteresis found for this phase transformation. Heat capacity measurements have indicated a similar transformation in $Ni(NO_3)_2 . 6NH_3$ (Long & Toettcher, 1946).

## 4    Spin state crossover in iron(II)
### 4.1    *Introduction*

One of the most interesting applications of Mössbauer spectroscopy to structural studies of iron(II) complexes is the investigation of spin state crossover, which is sometimes referred to as spin state transition or spin state equilibrium. This spin state crossover is possible because the iron(II) ion may adopt different ground state electronic configurations, depending upon the relative values of the crystal field potential $10Dq$ and the mean electron pairing energy $E_p$. Here $E_p$ is the energy required to spin pair the two additional electrons in the $t_{2g}$ orbitals. These electronic configurations are illustrated in Figure 3.23 for octahedral coordination geometry. It is found that in tetrahedral complexes $10Dq$ is significantly smaller than in octahedral complexes and is seldom of the order of the mean pairing energy. As a result, spin crossover is not found in tetrahedral complexes. If $10Dq^{oct}$ is much smaller than the mean pairing energy, Hund's rule will dictate the $t_{2g}^4 e_g^2$ configuration which has the $^5T_{2g}$ paramagnetic ground state of maximum spin multiplicity. In the opposite case, where $10Dq^{oct} > E_p$, the maximum multiplicity $^5T_{2g}$ state will be higher in energy than the spin paired diamagnetic $^1A_{1g}$ state arising from the ground state $t_{2g}^6 e_g^0$ configuration. By far the vast majority of the iron(II) complexes will fall into one or the other case, and either paramagnetic high spin $^5T_{2g}$, or diamagnetic low spin $^1A_{1g}$ complexes will result. All of the iron(II) compounds discussed so far fall into the weak field high spin configuration. In this discussion we will use the symmetry designations which apply to octahedral complexes. Small reductions from octahedral symmetry will in general not change the discussion of spin state crossover that follows.

In the special case in which the crystal field potential and the mean pairing energy are comparable, the $^5T_{2g}$ and $^1A_{1g}$ state may be quite close in energy. If their separation is of the order of $k_B T$, the thermal energy available to the molecule, a Boltzmann population of each state may produce a true temperature-dependent spin equilibrium between the two states. This is sometimes observed (Gütlich, 1981), but more often iron(II) exhibits spin crossover behaviour in which the change in spin state is accompanied by a structural transformation from a high spin structure above some crossover temperature to a low spin structure below that temperature. This structural transformation may in some instances be difficult to detect by crystallographic means, while in other cases distinct changes are observed (Ganguli, Gütlich, Muller & Irler, 1981). Often a long-range cooperative type of behaviour is theoretically expected and observed. There are now several hundred iron(II) compounds which exhibit spin crossover behaviour. Several recent review articles (e.g. Gütlich, 1981) describe these compounds in detail. Although in the subsequent discussion only one series of compounds will be covered it should be appreciated that a variety of complementary techniques and synthetic approaches are necessary to fully understand the structural and electronic properties of these compounds.

### 4.2 *Iron(II) phenanthroline complexes*

The iron(II) complexes of 1,10-phenanthroline and to a lesser extent $\alpha,\alpha'$-bipyridine form an ideal series to study because they span the

Fig. 3.23. The electronic configurations (*a*) and the spectroscopic states (*b*) arising in an octahedral iron(II) complex with differing crystal field parameters.

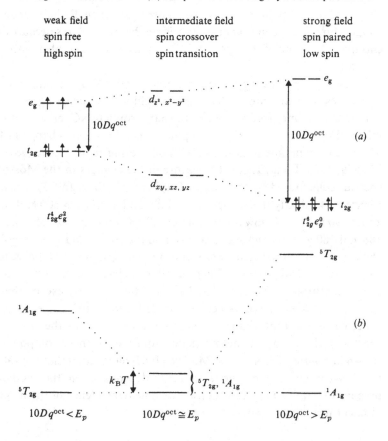

range of relative $10Dq^{oct}$ and $E_p$ values to produce high spin, low spin, and spin crossover compounds. Both the phenanthroline and bipyridine ligands are strong field ligands and it has long been known that their tris complexes are low spin and diamagnetic. By combining two of these ligands with other weak and strong field ligands, such as halides and cyanide, it has been possible to produce mixed ligand complexes which are high spin or low spin respectively. If intermediate field ligands such as thiocyanate $NCS^-$ or selenocyanate $NCSe^-$ are incorporated into the mixed ligand complex, the appropriate crystal field potential is generated to produce the spin crossover case. The Mössbauer parameters and the magnetic properties of several mixed ligand phenanthroline complexes (Berrett & Fitzsimmons, 1967; König, Hufner, Steichele & Madeja, 1967) are given in Table 3.8. It is clear that the complexes with the weak field weak $\sigma$-bonding halides are high spin $t_{2g}{}^4 e_g{}^2$ iron(II) compounds with large isomer shift values of around 1.0 mm/s and large quadrupole splittings in the range 2–3 mm/s, and with only a small change of these parameters with temperature. In contrast, the low spin complexes with the $t_{2g}^6$ configuration have much smaller isomer shifts and quadrupole splittings, which are again virtually independent of temperature. The small isomer shifts arise in these complexes because of the effective $\pi$-bonding of the $t_{2g}$ orbitals with the cyanide and phenanthroline ligands.

Much more interesting is the behaviour exhibited by the mixed ligand complexes with the intermediate ligand field pseudohalide ligands. Both $Fe(phenanthroline)_2(NCS)_2$ and $Fe(phenanthroline)_2(NCSe)_2$ as well as the related $Fe(bipyridine)_2(NCS)_2$ compound show a drastic change in their Mössbauer hyperfine parameters with decreasing temperature (König & Madeja, 1967; Long, 1968). The nature of the changes in the Mössbauer spectra is illustrated by the case of $Fe(phenanthroline)_2(NCS)_2$ shown in Figure 3.24. Clearly this compound is fully high spin from 300 K down to about 190 K, while below approximately 180 K it is fully low spin. Between 188 and 180 K both spin states coexist and the compound is undergoing a change in spin state or spin crossover from the $^5T_{2g}$ state above 190 K to the $^1A_{1g}$ state below 180 K. Very similar behaviour is observed in $Fe(phenanthroline)_2(NCSe)_2$ and $Fe(bipyridine)_2(NCS)_2$ except that the spin crossover observed is centred at 232 and 211 K respectively. As expected, the abrupt change in spin state observed in the Mössbauer spectra is also found in the magnetic properties of these compounds as shown in Figure 3.25. König & Madeja (1967) have found that the value of the low-temperature magnetic moment depends upon the method of preparation and some work indicates a time dependence of the magnetic moment near the crossover temperature.

Table 3.8. *Mössbauer parameters and magnetic properties of some iron(II) phenanthroline and related complexes*

| Compound | $T$/K | $\delta$/mm s$^{-1}$ [a] | $\Delta$/mm s$^{-1}$ | Effective magnetic moment/$\beta$ | $T$/K | Spin state |
|---|---|---|---|---|---|---|
| Fe(phenanthroline)$_2$Cl$_2$[b] | 295 | 1.01 | 2.94 | 5.20 | [c] | high spin |
| | 78 | 1.09 | 3.29 | 5.12 | | high spin |
| Fe(phenanthroline)$_2$Br$_2$[b] | 296 | 0.99 | 2.61 | 5.04 | [c] | high spin |
| | 78 | 1.10 | 3.23 | 4.93 | | high spin |
| Fe(phenanthroline)$_2$I$_2$[b] | 297 | 0.98 | 2.53 | 5.13 | [c] | high spin |
| | 78 | 1.06 | 3.13 | 4.86 | | high spin |
| Fe(phenanthroline)$_2$(N$_3$)$_2$[b] | 299 | 0.96 | 2.72 | 5.19 | [c] | high spin |
| | 78 | 1.07 | 3.06 | 5.07 | | high spin |
| Fe(phenanthroline)$_2$(NCS)$_2$[b, d] | 296 | 0.99 | 2.64 | 5.20 | 174 | high spin |
| | 78 | 0.44 | 0.31 | 1.47 | | low spin |
| Fe(phenanthroline)$_2$(NCSe)$_2$[b, d] | 299 | 0.99 | 2.44 | 5.04 | 232 | high spin |
| | 78 | 0.45 | 0.29 | 1.39 | | low spin |
| Fe(bipyridine)$_2$(NCS)$_2$[b] | 297 | 1.00 | 2.09 | 5.12 | 211 | high spin |
| | 78 | 0.42 | 0.53 | 1.25 | | low spin |
| Fe(phenanthroline)$_2$(NC)$_2$[e] | 298 | 0.18 | 0.58 | ~0 | [c] | low spin |
| Fe(phenanthroline)$_2$(NO$_2$)$_2$[f] | 293 | 0.28 | 0.38 | ~0 | [c] | low spin |
| | 78 | 0.25 | 0.41 | | | low spin |
| [Fe(phenanthroline)$_3$](ClO$_4$)$_2$[g] | 300 | 0.31 | 0.15 | ~0 | [c] | low spin |
| | 77 | 0.34 | 0.23 | ~0 | | low spin |

[a] Relative to iron metal.
[b] From Long (1968).
[c] No spin crossover is observed.
[d] From König & Madeja (1967).
[e] From Berrett & Fitzsimmons (1967).
[f] From König, Hufner, Steichele & Madeja (1967).
[g] From Bode, Gütlich & Koppen (1980), Collins, Pettit & Baker (1966) and Berrett, Fitzsimmons & Owusu (1968).

The cooperative spin crossover (as opposed to spin equilibrium) nature of the transition is revealed by the sharpness of the transition (see Figures 3.24 and 3.25), which cannot be fitted to a van Vleck expression for the magnetic moment of the two states (Figure 3.23) averaged over the

Fig. 3.24. The temperature dependence of the Mössbauer spectrum of Fe(phenanthroline)$_2$(NCS)$_2$, illustrating the spin crossover from high spin at 300 K to low spin at 77.4 K. (From Gütlich, 1981.)

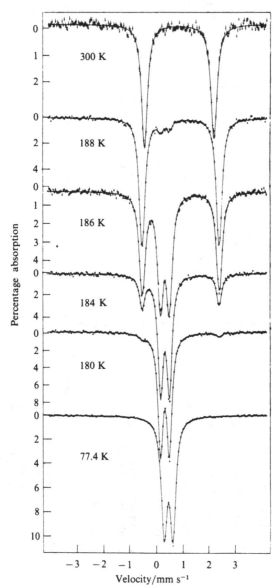

Boltzmann population of each state as a function of temperature (König & Madeja, 1967). This conclusion is further supported by recent X-ray powder diffraction studies on Fe(phenanthroline)$_2$(NCS)$_2$ which reveal different diffraction patterns for the two phases (Ganguli, Gütlich, Muller & Irler, 1981). The phase change associated with the transition is also indicated by a large endotherm in the differential thermal analysis seen at 178.6 K. An analogous change is observed in the infrared vibrational stretching frequency of the carbon–nitrogen bond in the thiocyanate ligand. As is seen in Figure 3.26, the room-temperature doublet observed in the infrared spectrum at 2060–2070 cm$^{-1}$ decreases on cooling and a new doublet appears at 2100–2110 cm$^{-1}$. The temperature dependence of these bands follows that observed for the magnetic moment and the 40 cm$^{-1}$ shift to higher energy is due to the changes in $\pi$-bonding between the $t_{2g}$ orbitals and the phenanthroline ligand upon change in spin state (König & Madeja, 1967). A further infrared study of the iron–ligand absorption bands has indicated a large increase in the iron–nitrogen bond order in passing from the high spin to the low spin state (Takemoto & Hutchinson, 1973). X-ray structural work has confirmed this conclusion for Fe(bipyridine)$_2$(NCS)$_2$, where it is found that upon cooling from 295 to 100 K the iron–nitrogen bond to the thiocyanate ligand is shortened by 0.008 nm while those to the bypyridine ligand decrease by 0.012 and 0.016 nm. All of these changes are consistent with a cooperatively induced phase transition and hence a spin crossover rather than a spin equilibrium situation.

Fig. 3.25. The temperature dependence of the effective magnetic moment $\mu_{\text{eff}}$ for three compounds undergoing spin state crossover.

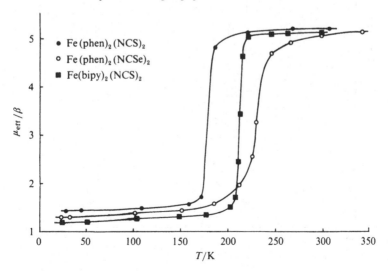

Probably the most important experimental work which confirms the cooperative nature of the spin crossover transition is the heat capacity measurements on $Fe(phenanthroline)_2(NCS)_2$ and $Fe(phenanthroline)_2$-$(NCSe)_2$ by Sorai & Seki (1974). The variation in the molar heat capacity of the latter compound with temperature is shown in Figure 3.27. Not only does the heat capacity show a sharp peak at the transition temperature (see Table 3.9), but there is also a change in the heat capacity, for the two different spin states (see Figure 3.27) at the transition. These authors propose that the total heat capacity of each spin state is made up of contributions from lattice vibrations, intramolecular vibrations, and electron thermal excitation. From this they have determined the changes in enthalpy and entropy associated with the change in spin state at the crossover. The results of these calculations are given in Table 3.9 and they

Fig. 3.26. The infrared spectrum of $Fe(phenanthroline)_2(NCS)_2$ obtained at (*a*) room temperature, (*b*) 172 K, and (*c*) 160 K.

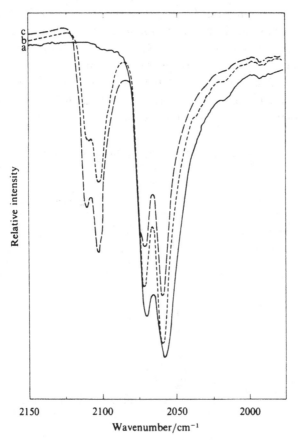

Table 3.9. *Thermodynamic changes at spin crossover*[a]

| Compound | $T_c$/K | $\Delta H$[b] | $\Delta S$[c] | $\Delta C_p$[d] |
|---|---|---|---|---|
| Fe(phenanthroline)$_2$(NCS)$_2$ | 176.29 | 8.6(1) | 48.8(7) | 18.7 |
| Fe(phenanthroline)$_2$(NCSe)$_2$ | 231.26 | 11.6(4) | 51(2) | 45.0 |

[a] Adapted from Sorai & Seki (1974).
[b] Change in the free energy in kJ mol$^{-1}$.
[c] Change in the entropy in J K$^{-1}$ mol$^{-1}$.
[d] Change in the heat capacity in J K$^{-1}$ mol$^{-1}$.

Fig. 3.27. The temperature dependence of the molar heat capacity $C_p$ for Fe(phenanthroline)$_2$(NCSe)$_2$. (From Sorai & Seki, 1974.)

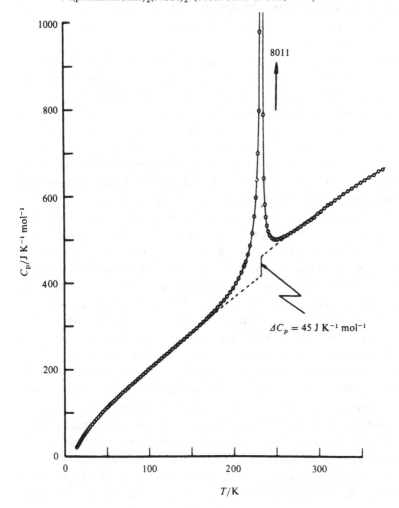

indicate that the major driving force for the spin transition is the large vibrational contribution to the entropy change at the transition. The thermal electronic excitations associated with the singlet to quintet spin state transition is only about 27% of the total entropy change. Of the remaining 73%, about half is associated with the iron–nitrogen vibrational changes and half with the nitrogen–iron–nitrogen vibrational deformation of the coordination environment. Apparently the lattice contribution to the entropy change at the transition is small. It is concluded that the cooperative nature of the spin state transition results because of the long-range correlation of the coupling of the electronic states to the phonon states of the crystal. This conclusion is supported by the Mössbauer study of the recoil-free fraction associated with the transition in $Fe(bipyridine)_2$ $(NCS)_2$. Whereas most physical properties of the transition show an abrupt change at or near the crossover temperature, the recoil-free fraction changes continuously.

These thermodynamic results have indicated that it might be possible to modify the cooperative interaction between the electronic spin state of the iron and the surrounding lattice by using metal dilution techniques. In an elegant series of experiments, Gütlich and his coworkers (Gütlich, 1981) have used this method to study a variety of spin crossover systems. In this context only those experiments dealing with the phenanthroline complexes will be discussed here. The technique involves the partial substitution of the iron(II) in the complex with a variety of similar divalent first row transition metal ions over a range of concentrations. If the divalent metal ion is larger than iron(II), as is the case for manganese(II), where the respective radii are 0.082 and 0.074 nm, then substitution of the iron(II) into $Mn(phenanthroline)_2(NCS)_2$ to yield a series of compounds $Fe_xMn_{1-x}(phenanthroline)_2(NCS)_2$ will place iron in a larger than expected lattice site. This may be thought of as a negative lattice pressure and will permit longer iron–ligand bonds, reducing the effective crystal field potential and favouring the high spin form of the complex. Thus, as x decreases from one, one would expect the high spin state to persist to a lower temperature at the expense of the low spin state, and thus the spin crossover temperature would be expected to decrease below the 174 K observed for the undiluted $Fe(phenanthroline)_2(NCS)_2$ complex. This is exactly what is observed, as is illustrated in Figure 3.28, for several compounds in which 0.1, 0.5 and 3% of iron are substitutionally doped into $Mn(phenanthroline)_2(NCS)_2$. At a fixed temperature of 5 K the relative amount of high spin iron(II) is found to increase dramatically as x decreases and the lattice pressure decreases. The influence of different doping levels on the value of the spin crossover temperature is shown in Figure 3.29, which indicates that as x decreases there is a substantial drop in the spin crossover temperature.

In contrast, if the divalent ion is smaller than iron(II), as is the case for cobalt(II) and nickel(II), which have ionic radii of 0.072 and 0.070 nm respectively, then substitution will place iron in a smaller than expected lattice site. In this case the positive lattice pressure will produce shorter iron–ligand bonds, and a stronger effective crystal field potential which will favour the low spin configuration and/or a higher spin crossover temperature. This is exactly what is observed in a series of

Fig. 3.28. The Mössbauer spectra of $Fe_xMn_{1-x}$(phenanthroline)$_2$(NCS)$_2$ obtained at 5 K and illustrating the increase in the high spin iron(II) fraction as the amount of iron decreases. (From Gütlich, 1981.)

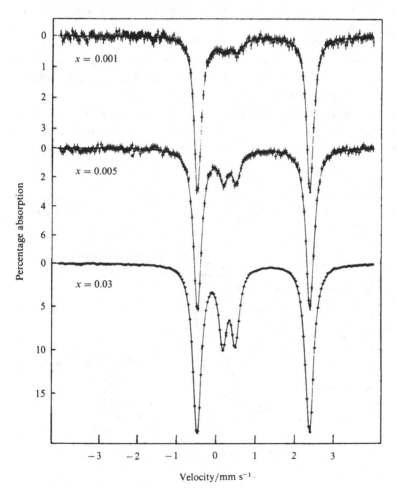

$Fe_xCo_{1-x}$(phenanthroline)$_2$(NCS)$_2$ complexes (Ganguli, Gütlich & Muller, 1982).

From the above it is apparent that Mössbauer spectroscopy, when combined with other techniques, provides a valuable means of following both structural and electronic changes occurring in a molecule or solid which is undergoing spin-state interconversion. As noted above, there are now several hundred such systems containing iron(II) and Mössbauer spectroscopy has proved to be useful in understanding the large majority of them (Gütlich, 1984).

### 4.3    *Light-induced excited state spin trapping*

Perhaps one of the most interesting recent applications of Mössbauer spectroscopy has been to the study of the absorption of radiation and the resultant spin state trapping in a compound exhibiting spin state crossover.

The iron(II) complexes having the formula $[FeL_6]X_2$, where L is an alkyltetrazole ligand, have been prepared and shown to exhibit spin state crossover (Franke, Haasnoot & Zuur, 1982). Müller, Ensling, Spiering & Gütlich (1983) have made a detailed study of the spin state crossover in several of these complexes where X is the $BF_4$ anion. The most interesting of these is $[Fe(ptz)_6](BF_4)_2$, where ptz is the 1-propyltetrazole ligand. This compound is found to be high spin at room

Fig. 3.29. The temperature dependence of the ratio of the high spin spectral area to the total spectral area in a series of $Fe_xMn_{1-x}$(phenanthroline)$_2$(NCS)$_2$ solid solutions. (From Gütlich, 1981.)

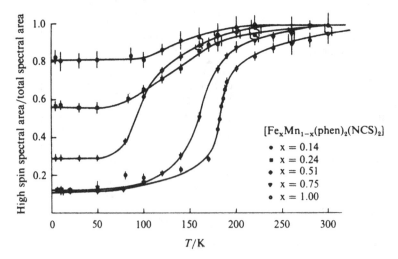

$[Fe_xMn_{1-x}(phen)_2(NCS)_2]$

- $x = 0.14$
- $x = 0.24$
- $x = 0.51$
- $x = 0.75$
- $x = 1.00$

temperature with a $^5T_{2g}$ ground state, and exhibits a room-temperature Mössbauer spectrum very similar to that shown in Figure 3.30(*b*). The asymmetry in the intensity of the two quadrupole split lines results from texture or crystallite orientation effects. Upon cooling to 128 K this compound undergoes a sharp spin state crossover, with a substantial temperature hysteresis, to the low spin state. Below about 97 K only a single absorption line is observed, similar to that shown in Figure 3.30(*a*). Very similar results are found for the same compound diluted with zinc and it has been concluded that a phase transition accompanies the spin crossover. All of the Mössbauer parameters, as well as the magnetic, optical and infrared data, are consistent with the spin crossover from a colourless high spin state to a deeply coloured low spin state.

Another area of interest is the behaviour of $[Fe(ptz)_6](BF_4)_2$ when it is irradiated with light from a xenon arc lamp (Decurtins *et al.*, 1984). The Mössbauer spectra of the irradiated samples are shown in Figure 3.30. The spectrum in Figure 3.30(*a*) was obtained at 15 K before irradiation and shows the singlet absorption line expected for this compound in its low spin state. The sample was then held at 15 K and irradiated for one hour with the xenon arc lamp and the subsequent Mössbauer spectrum obtained at 15 K (Figure 3.30(*b*)) reveals that irradiation has converted the sample to the high spin state. The Mössbauer parameters for the observed doublet are very similar to those found above 128 K for the unirradiated high spin sample and the results indicate that the light has completely converted the low spin complex to a high spin complex which has been trapped at 15 K. This phenomenon has been called light-induced excited spin state trapping (Decurtins *et al.*, 1984). The thermally trapped nature of the high spin state at 15 K may be demonstrated by warming the sample to different temperatures for various times. For example, warming the sample to 40 K for many hours had no noticeable influence on the sample, whereas warming to around 55 K for a few minutes and cooling to 15 K produced bleaching and eventually the elimination of the high spin component to yield a spectrum similar to Figure 3.30(*a*). Finally, upon warming the sample to 148 K and measuring the spectrum at this temperature the expected high spin compound is formed and the observed spectrum is very similar to Figure 3.30(*b*).

It has been proposed that the light-induced excited state spin trapping results from light excitation in a singlet to singlet absorption band at low temperature (Decurtins *et al.*, 1985). If there is a relaxation pathway from the excited spin singlet state to the excited $^5T_{2g}$ high spin state, and an energy barrier between this $^5T_{2g}$ high spin excited state and the $^1A_{1g}$ low spin ground state (see Figure 3.23), the molecule will be trapped into the metastable high spin excited state as indicated in Figure 3.30(*b*). As the temperature is increased, the thermal energy available to the system overcomes the energy barrier between the states and the system is bleached into the singlet ground

Fig. 3.30. The Mössbauer spectrum obtained at 15 K from [Fe(ptz)$_6$](BF$_4$)$_2$ measured (*a*) before exposure to light, (*b*) after exposure for one hour, and (*c*) after the sample had been warmed to 55 K for several minutes and recooled to 15 K. (From Decurtins *et al.*, 1984.)

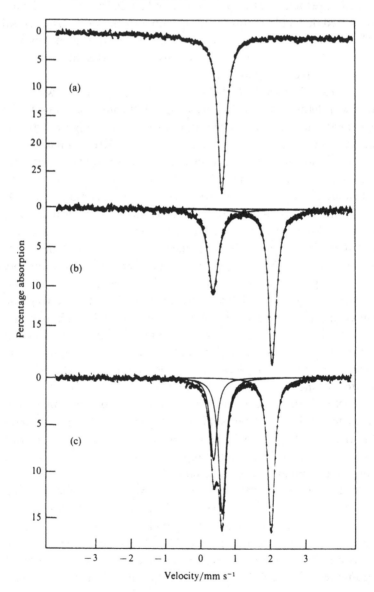

state. Similar preliminary experiments on $Fe(phenanthroline)_2(NCS)_2$ have indicated the presence of the same phenomenon.

## 5    Low spin iron(II) in iron pyrite

Strong crystal fields create $t_{2g}{}^6 e_g{}^0$ electronic configurations (see Figure 3.23) in which the equal occupation of the three $t_{2g}$ orbitals will lead to a zero valence contribution to the EFG as is indicated in Table 3.3. As a result the quadrupole splitting should be rather small and result predominantly from the lattice contribution to the EFG. It is thus quite surprising to find that iron pyrite $FeS_2$, a diamagnetic low spin iron(II) compound with the cubic Pa3 rock salt structure, exhibits a substantial positive quadrupole splitting of $+0.634(6)$ mm/s at room temperature (Montano & Seehra, 1976). The structure of $FeS_2$, illustrated in Figure 3.31, does show a trigonal distortion from octahedral symmetry which has been shown by Finklea, Cathey & Amma (1976) to be too small to produce a $(V_{zz})_{lat}$ term to account for the magnitude of the quadrupole splitting. It would appear therefore that the observed splitting must arise from a $(V_{zz})_{val}$ term derived from inequivalent occupation of the $t_{2g}$ orbitals. This inequivalence could arise from either the mixing of the $e_g$ orbitals into the $t_{2g}$ orbitals by the low symmetry $D_{3d}$ trigonal component to the crystal field or perhaps by a second-order spin–orbit interaction. The influence of the $D_{3d}$ trigonal field upon the octahedral crystal field orbitals is shown in Figure 3.32.

Fig. 3.31. The coordination geometry about the low spin iron(II) in $FeS_2$. (From Stevens *et al.*, 1980.)

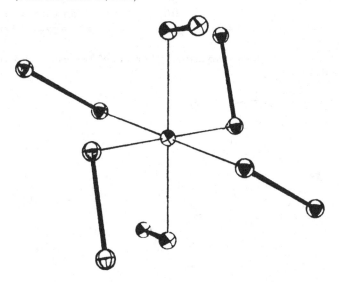

Stevens, DeLucia & Coppens (1980) have carried out a careful experimental single-crystal X-ray diffraction study of the electronic distribution in $FeS_2$ and have shown that the compound is low spin, but that there is a significant mixing of the orbitals as mentioned above. A population analysis in terms of the multipole density parameters has yielded the experimental populations of the $d$ orbitals given in Table 3.10, where the orbital symmetry designations are those given in Figure 3.32. It is immediately obvious that although the experimental populations are closest to those expected of a low spin compound there is a substantial occupancy of the $e_g'$, orbital derived from the $e_g$ excited state orbitals in octahedral symmetry. This occupancy indicates that there is a small but significant mixing of the orbitals by the low symmetry crystal field term. Following these considerations it becomes possible to calculate the electric field gradient, and hence the quadrupole splitting in $FeS_2$, by using the orbital populations given in Table 3.10. This yields the $V_{zz}$ value given in the table and a calculated quadrupole splitting of 0.8(5) mm/s. Although the X-

Table 3.10. *Iron orbital occupancies and electric field gradients in* $FeS_2$

| Orbital or EFG | Spherical atom occupancy | | Experimental occupancy |
|---|---|---|---|
| | High spin | Low spin | |
| $a_g$ | 2.0 | 2.0 | 2.0(1) |
| $e_g$ | 2.0 | 4.0 | 3.2(2) |
| $e_g'$ | 2.0 | 0.0 | 0.8(2) |
| $V_{zz}$ | $\gg 0$ | 0.0 | $-5.1 \times 10^{21}$ V/m$^2$ |

Fig. 3.32. The low symmetry crystal field splitting of the five octahedral 3$d$ orbitals in $FeS_2$.

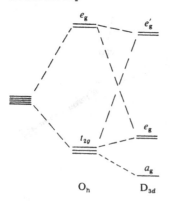

ray derived population densities do not lead to a very precise quadrupole splitting they do produce the proper magnitude and sign for $\Delta$. More important, it is possible to account for the large observed quadrupole splitting in terms of the $d$ orbital population.

It may at first seem surprising that the sulphide ligand should produce a crystal field potential sufficient to yield low spin compounds. However, high-pressure studies of both $^{57}$Fe substituted in $MnS_2$ (Bargeron, Avinor & Drickamer, 1971) and solid solutions of manganese and iron sulphide (Cheetham, Cole & Long, 1981) have revealed the presence of high spin iron(II) in the larger manganese sulphide lattice which would produce a negative lattice potential. The pressure studies indicate that a pressure of greater than 4 GPa is required to produce the low spin state for iron(II) in $MnS_2$.

## 6 High spin iron(III) complexes

### 6.1 *Introduction*

The $3d^5$ electronic configuration of iron(III) in the presence of a weak octahedral crystal field produces the $t_{2g}^3 e_g^2$ configuration with a $^6A_{1g}$ ground state as illustrated in Figure 3.33. The equivalent occupation of the five $3d$ orbitals gives, to a first approximation, a zero valence contribution to the electric field gradient. It would therefore be expected that octahedral iron(III) complexes will have small quadrupole splittings which arise mainly from the presence of a non-zero lattice contribution to the EFG. It should be noted in Figure 3.33 that small distortions from regular geometry, which do not change the sextet spin multiplicity of the iron(III)

Fig. 3.33. The orbital splitting diagram (*a*) and electronic ground state (*b*) for the high spin $3d^5$ iron(III) ion in an octahedral and a square pyramidal coordination geometry.

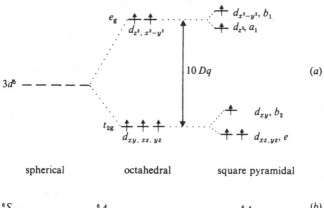

ion, have very little influence on $(V_{zz})_{val}$ and hence in high spin iron(III) compounds the quadrupole interaction does not provide a sensitive measure of the symmetry at the iron site. It is also found that high spin iron(III), as a result of the decreased shielding of the $s$ electron wavefunctions, has a lower isomer shift than is usually observed in iron(II) compounds. The typical values of the isomer shift of iron(III) are 0.3 to 0.6 mm/s, depending upon the nature of the coordination environment, significantly smaller values than the 0.9 to 1.3 mm/s range found for iron(III). As a result of these differences in isomer shift, Mössbauer spectroscopy can often be quite effective in distinguishing these different oxidation states and can therefore be helpful in studies of mixed valence iron-containing materials, as will be discussed later.

### 6.2     *Tetraphenylporphinato–iron(III) methoxide*

Despite the features discussed above concerning the small quadrupole interactions in high spin iron(III) it is frequently found that the observed quadrupole splittings are appreciably larger than can be accounted for by any reasonable lattice contribution to the EFG. One such case will now be considered in detail.

The Mössbauer spectra of a variety of iron(III) porphyrin complexes have been measured and found to exhibit quadrupole splittings within a surprisingly small range of values around 0.6 mm/s (Sams & Tsin, 1979). In several cases, including tetraphenylporphinato–iron(III) methoxide, $Fe(TPP)OCH_3$ (structure shown in Figure 3.34), the sign of $V_{zz}$ and hence of the quadrupole splitting has been shown to be positive (Dolphin, Sams, Tsin & Wong, 1978). The magnitude and sign of the quadrupole interactions in these compounds have been the subject of extensive discussion, usually based on the assumption of the equivalent occupation of the five $3d$ orbitals and a zero valence contribution to the EFG (see Table 3.11). The crystal field calculation of $(V_{zz})_{lat}$ yields a positive value of $V_{zz}$ (Harris, 1968), but in order to predict the magnitude of the quadrupole interactions it is necessary to assume a crystal field distortion which is very much larger than would be expected on the basis of the known structures. Unfortunately, simple molecular orbital calculations (Zerner, Gouterman & Kobayashi, 1966) give the wrong magnitude and sign for $(V_{zz})_{lat}$.

A recent detailed X-ray diffraction study of the experimental electron density distribution in $Fe(TTP)OCH_3$ has allowed the evaluation of the electron occupancies of the five $3d$ orbitals of the iron(III) ion (Lecomte, Chadwick, Coppens & Stevens, 1983). A multipole population analysis gave the results presented in Table 3.11 and indicated a net charge of $+2.5(2)$ on the iron ion, in good agreement with the expected oxidation

Table 3.11. *Iron orbital occupancies and electric field gradients in* Fe(TPP)OCH$_3$[a]

| Orbital or EFG | Spherical atom occupancy | Experimental occupancy |
|---|---|---|
| $b_1\ d_{x^2-y^2}$ | 1.00 | 1.04(6) |
| $a_1\ d_{z^2}$ | 1.00 | 1.07(6) |
| $b_2\ d_{xy}$ | 1.00 | 1.08(6) |
| $e\begin{cases} d_{xz} \\ d_{yz} \end{cases}$ | 1.00 | 0.90(8) |
| | 1.00 | 0.90(8) |
| $4s$ | 0.0 | 0.47(15) |
| $(V_{zz})_{val}$ | 0.0 | $+23 \times 10^{20}$ V/m$^2$ |
| $(V_{zz})_{lat}$ | — | $+1.05 \times 10^{20}$ V/m$^2$ |
| Calculated $\Delta$ | small | $\sim +0.6$ mm/s |
| Experimental $\Delta$ | — | $+0.56(5)$ mm/s |

[a] Adapted from Lecomte *et al.* (1983).

Fig. 3.34. The structure of Fe(TPP)OCH$_3$ obtained at 100 K and showing the eclipsed nature of the methoxide axial ligand and the approximate square pyramidal geometry at the iron site. (From Lecomte *et al.*, 1983.)

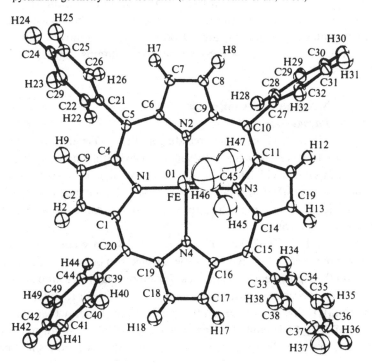

state. The charge on the methoxide ligand was $-0.6(2)$ and the charge on the pyrrole nitrogen atoms was $-0.85(2)$ each. The balance of charge was distributed on the TPP ligand. In this calculation of the orbital populations, the non-integral populations were associated with a mixing into the ground state of other low-lying excited states and with metal ligand covalency. The populations are in excellent agreement with those expected of a high spin iron(III) compound. The subsequent calculation of $(V_{zz})_{val}$ and $(V_{zz})_{lat}$ based on these populations and the spatial charge distribution within the ligands shows that the major contribution to the EFG results from $(V_{zz})_{val}$ and the inequivalence of the orbital populations. The resulting calculated quadrupole splitting of about $+0.6$ mm/s is in excellent agreement with the observed value of $+0.56(5)$ mm/s (Dolphin *et al.*, 1978).

The orbital populations observed for $Fe(TPP)OCH_3$ agree very well with those found for Co(TPP) (Stevens, 1981) and indicate that the deformation densities may be quite similar among structurally similar metalloporphyrins. If this is the case the results and conclusions obtained for $Fe(TTP)OCH_3$ may apply to many related porphyrin iron complexes with similar quadrupole interactions. Overall these results indicate that the orbital populations predicted by simple crystal field theory are quite accurate in describing the electronic configuration of the iron. However, a detailed evaluation of the relative contributions of $(V_{zz})_{val}$ and $(V_{zz})_{lat}$ to the observed quadrupole interaction requires more accurate, and experimentally determined, orbital populations. To date, similar results have been obtained for $KFeS_2$ which also contains high spin iron(III) (Stevens, 1982).

# 7    Mixed valence compounds
## 7.1    *Introduction*

The discussion in the preceding sections has been concerned with compounds which show a single valence or oxidation state. In this section some compounds which have iron in more than one oxidation state, usually iron(II) and iron(III), will be considered. Mössbauer spectroscopy can be very useful in determining the structural and electronic properties of such mixed valence compounds.

Robin & Day (1967) have classified mixed valence compounds on the basis of the energy required to transfer an electron between the low valence site and the high valence site. In Class I compounds, the ground state M(II)/M'(III) electronic configuration (where M and M' refer to different iron(II) and iron(III) sites in a compound) is much lower in energy than the excited state M(III)/M'(II) configuration which may result from an intervalence electron transfer. In this type of compound the two iron sites may be chemically and structurally quite different and are probably well isolated from each other. The compound would then be expected to exhibit physical properties corresponding to a superposition of the two different valence states.

Class II compounds contain M and M' sites which, although crystallographically distinct, are chemically similar. They might for instance both have the same coordination geometry but with different bond lengths or perhaps slightly different coordination environments. In this case the best electronic ground state wavefunction is a linear combination consisting mainly of the M(II)/M'(III) electronic configuration with some contribution from the M(III)/M'(II) configuration. The extent of mixing of the M(III)/M'(II) configuration with the M(II)/M'(III) configuration will depend upon $\alpha$, the electron delocalisation coefficient. If $\alpha$ is zero, Class I behaviour is observed. If $\alpha$ is small, but non-zero, then there is a perturbation matrix element which mixes the two configurations. If the energy between the two states of different configurations is of the order of a few hundred kJ/mol, then an intervalence charge transfer electronic absorption band will be present, often in the optical portion of the spectrum. It is this charge transfer band which gives many mixed valence compounds their intense colours. However, since $\alpha$ is still small, any electronic excitations localised on either M(II) or M'(III) will be similar to those found in the respective single valence compounds. In some cases the valence electron transfer process might occur, perhaps by thermal activation, on the measurement timescale associated with a specific physical measurement. In this case the results will be indicative of an averaged valence on this timescale and will most likely be highly temperature dependent. Hence, depending on the type of measurement undertaken, the Class II compounds might show discrete or averaged valence characteristics.

Class III compounds may have a stoichiometry indicative of a non-integral valence for the metal M, and structures which do not contain crystallographically distinct sites for the localisation of the different valence states. In this case the mixed valence electron delocalisation coefficient $\alpha$ will be close to 0.7 and the ground state wavefunction must be an approximately equal combination of the M(II)/M'(III) and M(III)/M'(II) configurations, where M and M' are essentially indistinguishable. Class III compounds will have an energy for intervalence charge transfer which is virtually zero and hence will often show conductivity, optical and magnetic properties typical of metals if the compounds have a long-range or three-dimensional lattice. In contrast, if the materials are not infinite lattices but contain discrete clusters or polynuclear molecules, they may not show metallic properties, but may exhibit properties that are not characteristic of those found in related materials that do not have mixed valence.

As mentioned above, the results obtained in the study of a mixed valence compound depend upon the characteristic time associated with the measurement. For Mössbauer spectroscopy an upper limit on this characteristic time is given by the mean lifetime of the nuclear excited state involved in the Mössbauer transition, which in the case of $^{57}Fe$ is 97.8 nanoseconds (see Chapter 5 for a more complete discussion of the timescales associated with a Mössbauer spectroscopy measurement). The Mössbauer spectral parameters will depend upon the typical electronic environment that the nucleus senses during this characteristic measurement time.

Roughly three spectral regimes may be observed (Brown & Wrobleski, 1980). In the first (i), which is typically associated with Class I and some Class II compounds, the intervalence charge transfer rate is very slow relative to the Mössbauer timescale and the nucleus effectively senses one electronic valence state. Hence the Mössbauer spectrum will be a superposition of that expected for each separate valence state, giving discrete iron(II) and iron(III) subspectra. In the second regime (ii), the timescale of the intervalence charge transfer and the Mössbauer timescale are comparable and depending upon the temperature and/or pressure either discrete or averaged electronic valence states may be observed in the Mössbauer spectrum. It is in general more likely to see the averaged valence state at higher temperatures where the thermal activation of the intervalence charge transfer will be most significant. Unfortunately many potential compounds of this type cannot sustain the high temperatures needed for this thermal activation.

In the third regime (iii), the rate of intervalence charge transfer is rapid on the Mössbauer timescale and the Mössbauer spectrum will exhibit essentially the average of the different valence state spectra. In this case the observed Mössbauer hyperfine parameters are roughly the average of those found in structurally similar but single valence iron(II) and iron(III) compounds. In the following discussion some compounds which fit into each of these spectral regimes and mixed valence classes will be considered.

### 7.2    *Mixed valence iron fluoride hydrates*

The Mössbauer parameters of $Fe_2F_5 \cdot 7H_2O$ and $Fe_2F_5 \cdot 2H_2O$ are presented in Table 3.12 and the room temperature spectrum of $Fe_2F_5 \cdot 7H_2O$ is given in Figure 3.35. The Mössbauer parameters for the two outer lines observed in the spectrum of $Fe_2F_5 \cdot 7H_2O$ can be assigned to the discrete iron(II) valence species $[Fe(H_2O)_6]^{2+}$ (Walton, Corvan, Brown &

Table 3.12. Mössbauer parameters of some mixed valence compounds

| Compound | T/K | $\delta$/mm s$^{-1}$ [a] | $\Delta$/mm s$^{-1}$ | Species | Class[b] | Regime[b] |
|---|---|---|---|---|---|---|
| Fe$_2$F$_5\cdot$7H$_2$O[c] | 300 | 1.25 | 3.31 | [Fe(II)(H$_2$O)$_6$] | I | i |
| | | 0.44 | 0.59 | [Fe(III)(H$_2$O)F$_5$] | | i |
| Fe$_2$F$_5\cdot$2H$_2$O[d] | 295 | 1.34 | 2.44 | [Fe(II)(H$_2$O)$_2$F$_4$] | II | i |
| | | 0.44 | 0.65 | [Fe(III)F$_6$] | | i |
| | 55 | 1.45 | 3.28 | [Fe(II)(H$_2$O)$_2$F$_4$] | II | i |
| | | 0.55 | 0.58 | [Fe(III)F$_6$] | | i |
| Prussian Blue | 298[e] | −0.04 | ~0 | [Fe(II)(CN)$_6$] | II | i |
| | | 0.46 | 0.47 | [Fe(III)(NC)$_{4.5}$(OH$_2$)$_{1.5}$] | | i |
| | 78[f] | −0.08 | ~0 | [Fe(II)(CN)$_6$] | II | i |
| | | 0.49 | 0.57 | [Fe(III)(NC)$_{4.5}$(OH$_2$)$_{1.5}$] | | i |
| Biferrocene[g] | 78 | 0.52 | 2.36 | [$\eta^5$-C$_5$H$_5$)Fe(II)($\eta^5$-C$_5$H$_4$)] | h | — |
| Biferrocene$^{1+}$[g] | 298 | 0.419 | 2.053 | [$\eta^5$-C$_5$H$_5$)Fe(II)($\eta^5$-C$_5$H$_4$)] | I | i |
| | | 0.421 | 0.302 | [$\eta^5$-C$_5$H$_5$)Fe(III)($\eta^5$-C$_5$H$_4$)] | | |
| | 78 | 0.510 | 2.141 | [$\eta^5$-C$_5$H$_5$)Fe(II)($\eta^5$-C$_5$H$_4$)] | | |
| | | 0.518 | 0.288 | [$\eta^5$-C$_5$H$_5$)Fe(III)($\eta^5$-C$_5$H$_4$)] | | |
| Biferrocene$^{2+}$[g, i] | 78 | 0.497 | 0.163 | [$\eta^5$-C$_5$H$_5$)Fe(III)($\eta^5$-C$_5$H$_4$)] | h | — |
| Biferrocenylene[i] | 300 | 0.434 | 2.397 | [$\eta^5$-C$_5$H$_4$)$_2$Fe(II)]$_2$ | h | — |
| | 4.2 | 0.531 | 2.405 | [$\eta^5$-C$_5$H$_4$)$_2$Fe(II)]$_2$ | | |
| Biferrocenylene$^{1+}$[i] | 300 | 0.441 | 1.719 | [$\eta^5$-C$_5$H$_4$)$_2$Fe(II, III)]$_2$ | III | iii |
| | 4.2 | 0.542 | 1.756 | [$\eta^5$-C$_5$H$_4$)$_2$Fe(II, III)]$_2$ | | |
| Ilvaite[j] | 190 | 0.98 | 2.23 | [Fe$_B$(II)O$_6$] | II | i |
| | | 0.86 | 2.50 | [Fe$_A$(II)O$_6$] | | i |
| | | 0.30 | 1.40 | [Fe$_A$(III)O$_6$] | | i |
| | 410 | 0.84 | 2.18 | [Fe$_A$(II, III)O$_6$] | II | iii |
| | | 0.45 | 1.40 | [Fe$_B$(II)O$_6$] | | i |
| | 610 | 0.73 | 1.95 | [Fe$_B$(II)O$_6$] | II | i |
| | | 0.35 | 1.20 | [Fe$_A$(II, III)O$_6$] | | iii |

[a] Relative to iron metal.
[b] See text.
[c] From Brown & Wrobleski (1980), and Walton et al. (1976).
[d] From Walton et al. (1977).
[e] From Bonnette & Allen (1971).
[f] From Maer et al. (1968).
[g] From Cowan et al. (1971).
[h] Single valent compound included for comparison.
[i] From Morrison & Hendrickson (1975).
[j] From Heilmann et al. (1977).

Day, 1976). A comparison of the parameters given in Table 3.12 with those for the $[Fe(H_2O)_6]^{2+}$ ion found in $Fe(BF_4)_2 . 6H_2O$ and $Fe(ClO_4)_2 . 6H_2O$ (Table 3.7) reveals a strong similarity and supports this assignment. The inner lines, whose spectral parameters are typical of discrete iron(III), are assigned to the $[Fe(H_2O)F_5]^{2-}$ moiety containing high spin iron(III). The absorption areas associated with each valence state are in a ratio of one to one, indicating that the recoil-free fractions are equivalent for each site, an equivalence that need not always be the case. Related broad line $^{19}F$ nuclear magnetic resonance and X-ray powder diffraction results (Sakai & Tominaga, 1975) have supported these structural conclusions. The Mössbauer data and the absence of any intervalence charge transfer absorption band in the electronic spectrum of $Fe_2F_7 . 7H_2O$ indicate that it is a Class I mixed valence material with well-trapped valence states. A low-temperature magnetic study of this material has revealed an antiferromagnetic transition at close to 3 K (Jones, Hendricks, Auel & Amma, 1977).

The $Fe_2F_5 . 2H_2O$ compound has a structure of vertex-sharing octahedra of iron(III) $(FeF_6)$ units forming parallel zig-zag chains which are cross linked by *trans* $[Fe(H_2O)_2F_4]$ units containing iron(II). A schematic drawing of an isolated fragment of this three-dimensional network is shown

Fig. 3.35. The room temperature Mössbauer spectrum of $Fe_2F_5 \cdot 7H_2O$, showing the discrete valence spectra of iron(II), the outer lines, and iron(III), the inner lines. (From Walton *et al.*, 1976.)

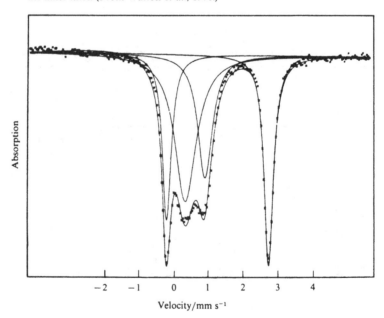

Velocity/mm s$^{-1}$

in Figure 3.36. Once again the Mössbauer parameters (Table 3.12) reveal that on the Mössbauer timescale $Fe_2F_5 . 2H_2O$ has discrete valence states. However, the presence of a strong optical absorption band at $15\,000\,cm^{-1}$, associated with intervalence charge transfer, indicates that this compound is a Class II mixed valence compound, while magnetic measurements indicate that $Fe_2F_5 . 2H_2O$ orders ferromagnetically at 48 K as a result of spin compensation of the iron(III) by the iron(II) ions (Walton, Brown, Wong & Reiff, 1977).

### 7.3    *Prussian Blue:* $Fe_4[Fe(CN)_6]_3 . 14H_2O$

One of the most well-known mixed valence compounds is the pigment Prussian Blue which is a Class II compound. There has been considerable controversy as to whether Prussian Blue, which may be produced by the reaction,

$$2Fe_2(SO_4)_3 + 3K_4Fe(CN)_6 \rightarrow Fe(III)_4[Fe(II)(CN)_6]_3 + 6K_2SO_4,$$

and the related compound, Turnbull's Blue, produced by the reaction,

$$FeCl_3 + 3FeCl_2 + 3K_3Fe(CN)_6 \rightarrow Fe(III)_4[Fe(II)(CN)_6]_3 + 9KCl,$$

are the same or not. It has been possible to show by the use of Mössbauer spectroscopy that these two preparations give the same compound and that Prussian Blue and Turnbull's Blue are identical (Maer, Beasley, Collins & Milligan, 1968; Bonnette & Allen, 1971).

The Mössbauer spectrum of Prussian Blue is very complex and the best understanding of its deconvolution into component spectra may be obtained by studying compounds selectively labelled with $^{56}Fe$ and the Mössbauer isotope $^{57}Fe$. Bonnette & Allen (1971) have been able to obtain $^{56}Fe(III)_4[Fe(II)(CN)_6]_3 . xH_2O$,    $Fe(III)_4[^{56}Fe(II)CN_6]_3 . xH_2O$    and

Fig. 3.36. An isolated fragment of the three-dimensional structure of $Fe_2F_5.2H_2O$. (From Hall *et al.*, 1977.)

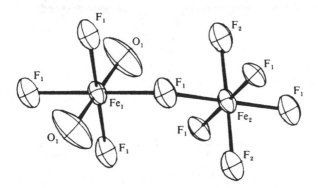

$^{57}Fe(III)^{56}Fe(III)_3[^{56}Fe(II)(CN)_6]_3 . xH_2O$, where Fe represents natural abundance iron. The resulting Mössbauer spectral assignments are given in Table 3.12 and the spectra are shown in Figure 3.37. These results indicate that Prussian Blue contains low spin iron(II) in the $[Fe(CN)_6]^{4-}$ moiety in which the iron(II) is octahedrally coordinated to six carbon atoms of the cyanide ligands, as illustrated in Figure 3.38(a). The iron(III) is present in

Fig. 3.37. The Mössbauer spectra obtained at 298 K from: (a) $^{56}Fe(III)_4[Fe(II)(CN)_6]_3 . xH_2O$, (b) $Fe(III)_4[^{56}Fe(II)(CN)_6]_3 . xH_2O$, and (c) $^{57}Fe(III)^{56}Fe(III)_3[^{56}Fe(II)(CN)_6]_3 . xH_2O$.

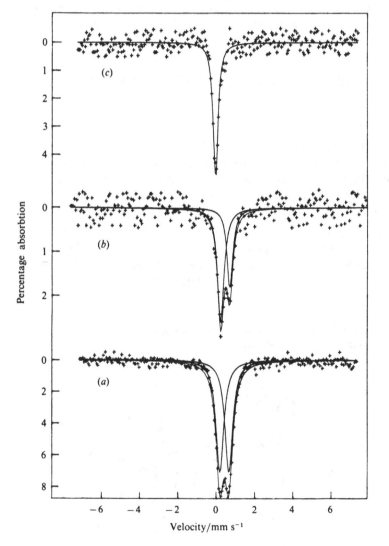

the high spin state and has, as a result of the disordered structure of Prussian Blue (see below), a coordination geometry on average of $[Fe(CN)_{4.5}(OH_2)_{1.5}]^{1.5-}$. The intervalence charge transfer rate is slow relative to the Mössbauer timescale and Mössbauer spectra representing the discrete low spin iron(II) and high spin iron(III) valence states are observed. However, the intense colour of Prussian Blue indicates that it is a Class II mixed valence compound.

The controversy concerning the structure of Prussian and Turnbull's Blue has partly resulted from the difficulty of producing good single crystals. Neutron diffraction work by Ludi and his coworkers (Buser, Schwarzenbach, Petter & Ludi, 1977; Herren, Fischer, Ludi & Hälg, 1980) has shown that Prussian Blue has a face centred cubic structure with the Fm3m space group. This leads to the averaged ideal structure shown in Figure 3.38(a) in which each metal site is occupied in an ordered fashion by iron(II) or iron(III) and cyanide bridges link each iron site to produce octahedral symmetry. The actual structure is more complex because the true stoichiometry, $Fe(III)_4[Fe(II)(CN)_6]_3 \cdot 14H_2O$, indicates that iron(II) and iron(III) are not present in the one-to-one ratio required by the ideal face centred cubic structure. In order to compensate for the deficiency of iron(II), Prussian Blue adopts a structure in which vacancies of $[Fe(II)(CN)_6]$ occur, as illustrated in Figure 3.38(b). These vacancies are filled with six interstitial water molecules that are coordinated to the iron(III) ions. Thus, of the 14 water molecules typically found in Prussian

Fig. 3.38. (a) The ideal face centred cubic structure of Prussian Blue, where open points represent iron(II) and closed points iron(III). Each line represents a cyanide ligand with the carbon bonded to the iron(II) and the nitrogen to iron(III). (b) The structure of Prussian Blue showing the disordered vacancies resulting from a missing $[Fe(II)(CN)_6]$ unit. The large open circles represent the coordinated oxygen of water and the large closed points the hydrogen bonded uncoordinated water.

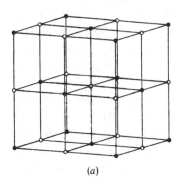

(a)

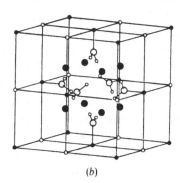

(b)

Blue, six are coordinated to iron(III) and eight are either isolated in the centre of the cell octants or hydrogen bonded to the coordinated water in the vacancies. The location of the $[Fe(II)(CN)_6]$ vacancies is disordered such that the averaged structure observed in a diffraction experiment has face centred cubic symmetry. On average the iron(III) site has a coordination environment with 4.5 cyanide ligands and 1.5 water molecules as indicated above and in Table 3.12. Actually, particular iron(III) ions have a range of different coordination environments and the resulting Mössbauer spectrum represents an average of these (Herren *et al.*, 1980).

### 7.4    *Mixed valence ferrocene derivatives*

The various derivatives of ferrocene $[(\eta^5\text{-}C_5H_5)_2Fe]$ are ideal candidates for mixed valence interactions because of the extensive modifications to the electronic structure of the iron which can be produced by changing the organic ligands. It has been known for some time that, although ferrocene and the ferrocenium cation $[(\eta^5\text{-}C_5H_5)_2Fe]^+$ produce very similar isomer shifts, their quadrupole splittings are quite different at 2.42 mm/s, and between zero and 0.1 mm/s respectively. Thus, the quadrupole interaction can be used as a measure of the mixed valence interaction. This is indicated by the results given in Table 3.12 for biferrocene and its mono and dicationic complexes. Biferrocene, in which each iron is formally low spin iron(II), has a substantial quadrupole splitting of 2.36 mm/s at 78 K, whereas the dication of biferrocene in which each iron is formally iron(III), has a quadrupole splitting of only 0.16 mm/s (Cowan, Collins & Kaufman, 1971). As might be expected, the monocation of biferrocene with iron(II) and iron(III) is a Class I mixed valence material and shows a Mössbauer spectrum that is basically a superposition of the two valence states (see Table 3.12).

In contrast, in biferrocenylene and its mono and dicationic complexes, different behaviour is observed (Morrison & Hendrickson, 1975). The single valence biferrocenylene compound gives results quite similar to ferrocene and biferrocene, whereas the monocation, with formally one iron(II) and one iron(III) ion, shows only a single quadrupole split doublet with a splitting of only 1.72 mm/s at 300 K. Virtually the same result is obtained at 4.2 K and the intermediate quadrupole splitting is indicative of a mixed valence compound in which the intervalence electron transfer rate is fast on the Mössbauer timescale. This and related properties of this monocationic complex indicate that it is a Class III mixed valence compound. Rather surprisingly, the dication of biferrocenylene shows a large quadrupole splitting rather than the small value which might be expected. An applied field study indicated that the quadrupole interaction is negative, in contrast to the positive value obtained for ferrocene (Collins, 1965). A variety of other related, but more complex structures give similar results and indicate that the proximity of the iron ions, and the number and nature of the exchange pathways connecting the ions, are very important in determining the rate of intervalence electron transfer.

### 7.5    *Ilvaite: a mixed valence mineral*

A number of minerals exhibit mixed valence behaviour, of which the best known is probably magnetite, the ferrospinel Fe(III)-[Fe(II)Fe(III)]$O_4$, which has been extensively studied by Mössbauer spectroscopy and other techniques. A detailed discussion of this complex mineral has been given by Goodenough (1980) and here a simpler example, the mixed valence iron silicate mineral ilvaite, $CaFe(II)_2Fe(III)Si_2O_8(OH)$, will be considered. The structure of ilvaite, illustrated in Figure 3.39(*a*), contains two iron sites Fe$_A$ and Fe$_B$ and consists of infinite edge-sharing double chains of Fe$_A$, with Fe$_A$ to Fe$_A$ distances of 0.303 and 0.283 nm within the double chain (see Figure 3.39(*b*)) (Yamanaka & Takeuchi, 1979). The Fe$_B$ sites are bound to these chains at a distance of 0.315 and 0.325 nm from the Fe$_A$ sites. The neutron diffraction results indicate that only iron(II) is found on the Fe$_B$ sites, whereas equal amounts of iron(II) and iron(III) may be found on the Fe$_A$ sites. The structure thus provides a variety of structural pathways for intervalence charge transfer.

The Mössbauer spectra of ilvaite obtained between 128 and 898 K are illustrated in Figure 3.40 (Heilmann, Olsen & Olsen, 1975). A smooth uniform variation in the spectrum in the 1 to 2 mm/s region is observed between these temperatures. Grandjean & Gerard (1975) have attributed the spectral variation to an increase in iron(III) character of one of the iron(II) ions as it gives up an electron through intervalence charge transfer The analysis of the spectra shown in Figure 3.40 by Heilmann *et al.* (1975) supports this view and indicates a smooth variation in the Mössbauer hyperfine parameters for ilvaite as a function of temperature. The variation in the isomer shift is shown in Figure 3.41 and typical values and assignments are given in Table 3.12.

Ilvaite is an especially interesting mixed valence mineral because it shows one iron(II) site, Fe$_B$, which retains a unique valence up to at least 900 K. Apparently the transfer of an electron from Fe$_B$ across the shared face to Fe$_A$ is difficult in comparison with transfer between the edge-shared Fe$_A$ sites. The Mössbauer parameters for the Fe$_B$ site given in Table 3.12 and the temperature dependence of its isomer shift (see Figure 3.41) indicate that it remains high spin iron(II). The situation for the Fe$_A$ atoms in the double chain is quite different, since below about 300 K the Mössbauer spectrum

Fig. 3.39. (a) The structure of ilvaite illustrating the edge-sharing double chains of $Fe_A$ which extend along the crystallographic $c$ axis, and the more isolated six coordinate $Fe_B$. (From Yamanaka & Takeuchi, 1979.) (b) A projection of the $Fe_A$ double chain in ilvaite, O(7) is the hydroxyl group. Interionic distances (in nm) for the various $Fe_A$–$Fe_A$ and $Fe_A$–$Fe_B$ interaction pathways are also shown. (From Nolet & Burns, 1979.)

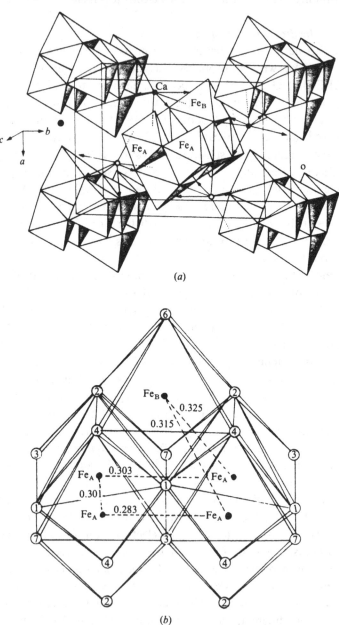

(a)

(b)

can be resolved into three quadrupole split doublets. One can be assigned to the $Fe_B$ site, while the second, with an isomer shift of about 0.9 mm/s and a quadrupole splitting of about 2.5 mm/s, can be assigned to $Fe_A$ in the divalent state. The third doublet, with the smallest isomer shift and quadrupole splitting (see Table 3.12), may be assigned to $Fe_A$ in the trivalent state. Between about 300 and 400 K the absorption lines associated with the second and third doublet coalesce to give a single doublet above 400 K (see Figures 3.40 and 3.41) with hyperfine parameters that are roughly the average of those expected for the discrete valence states. This behaviour is indicative of electron exchange or intervalence electron transfer along or between the $Fe_A$ double chains. Mössbauer spectroscopy is ideal for

Fig. 3.40. The Mössbauer spectra of ilvaite. (From Heilmann *et al.*, 1977.)

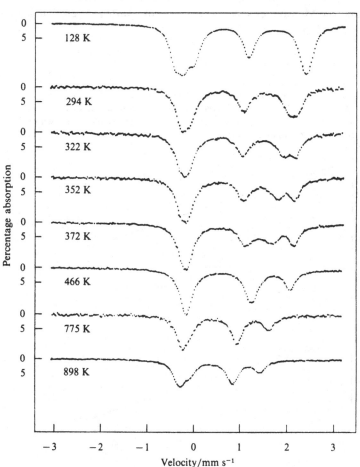

demonstrating this change, because between about 300 and 400 K the rate of electron exchange between the $Fe_A$ sites changes from being slow on the Mössbauer timescale to being fast on this timescale. Hence ilvaite exhibits a fixed valence for the $Fe_B$ sites and a mixed valence interaction for the $Fe_A$ sites for which all three regimes can be observed in the Mössbauer spectra. More recently, Litterst & Amthauer (1984) have confirmed the earlier work and have shown that by using a relaxation model for stochastic electronic valence state fluctuations for the iron(II) and iron(III) ions on the $Fe_A$ sites it is possible to calculate the activation energy for this process. The fluctuation rate follows an Arrhenius law temperature dependence, with an activation energy of 11.6 kJ/mol, and shows non-adiabatic electron hopping. This model produces an excellent fit to the experimental Mössbauer spectra, as shown by the solid line in Figure 3.42. In some related work, Nolet & Burns (1979) and Yamanaka & Takeuchi (1979) have shown that the Mössbauer spectra observed for ilvaite are very sensitive to the presence of impurities on the metal sites.

Fig. 3.41. (*a*) The temperature dependence of the isomer shifts of the quadrupole split doublets observed in ilvaite. The mean value of the isomer shifts for the iron(II)–iron(III) temperature-sensitive doublets is shown in (*b*). (From Heilmann *et al.*, 1977.)

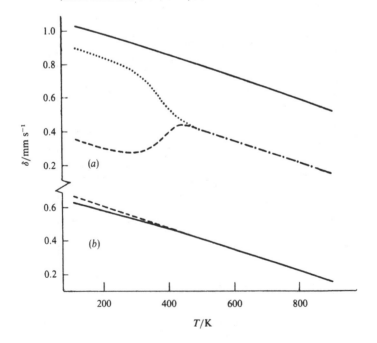

## 8 Dynamic structural studies

In the previous section, the ability of Mössbauer spectroscopy to measure the electron hopping or exchange rate between iron sites of different valence was discussed in detail. In this section we will consider a rather different dynamic effect which is amenable to study by Mössbauer spectroscopy. In all of the above discussions it has been assumed that the Mössbauer absorber nucleus is embedded in a relatively static molecular framework in which each atom is at most undergoing a vibration about its equilibrium lattice position. This situation will often lead to a well-defined electric field gradient tensor that changes only slightly with temperature in the absence of any crystallographic phase changes. In contrast, if the Mössbauer absorber nucleus is placed in a molecule which is free to undergo dynamic changes in its environment, such as molecular rotation

Fig. 3.42. Mössbauer spectra of ilvaite obtained between 298 and 455 K. The solid line fits to the data represent a stochastic relaxation model composed of the quadrupole split doublets discussed in the text. (From Litterst & Amthauer, 1984.)

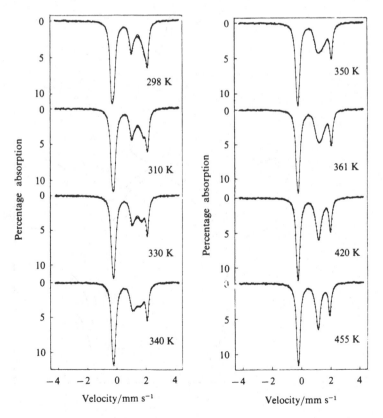

138     *Mössbauer spectroscopy as a structural probe*

about a fixed point in the lattice, then the situation may be quite different. If the molecular site symmetry at the absorber nucleus is less than cubic the quadrupole interaction may be time dependent as a result of the fluctuating electric field gradient. This fluctuating electric field gradient is usually described in terms of the relaxation time required for the electric field gradient to relax from one orientation to another (Hoy, 1984). If this relaxation is isotropic only a single relaxation time is sufficient to describe the relaxation process and Mössbauer spectra of the type shown in Figure 3.43 will be observed, depending upon the relaxation time (Fitzsimmons,

Fig. 3.43. The influence of an isotropic fluctuation in the electric field gradient on the Mössbauer spectrum. The relaxation time (in seconds) is given next to each spectrum. (From Fitzsimmons & Hume, 1980.)

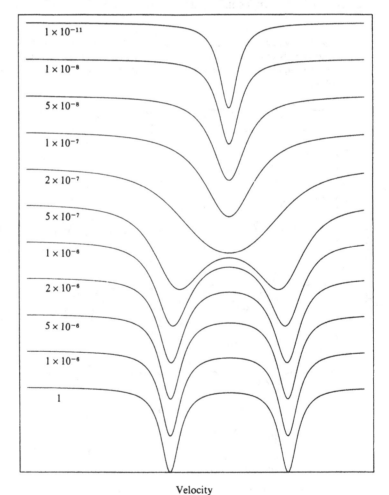

Velocity

1980; Fitzsimmons & Hume, 1980). If the relaxation time is long (say, $10^{-5}$ s) relative to the lifetime of the Mössbauer nuclear state, then the nucleus experiences a quadrupole interaction and a doublet is observed, as illustrated at the bottom of Figure 3.43. As the relaxation time shortens, the quadrupole interaction decreases until at very short relaxation times (say, $10^{-11}$ s) the electric field gradient averages to zero over the lifetime of the nuclear state and a single line is observed in the Mössbauer spectrum. The situation may be more complex if the relaxation of the electric field gradient is anisotropic, as is illustrated in Figure 3.44. In this case more than one relaxation time may be required to describe the relaxation process and the

Fig. 3.44. The influence of an axial anisotropic fluctuation in the electric field gradient on the Mössbauer spectrum. For these calculations the equatorial relaxation time is assumed to be long and the axial relaxation times (in seconds) are shown on the diagram. (From Fitzsimmons & Hume, 1980.)

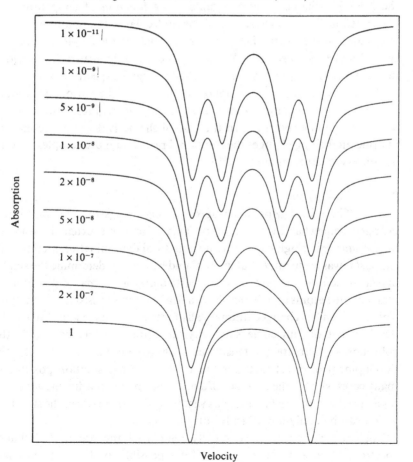

quadrupole interaction need not necessarily average to zero for rapid relaxation in only one orientation.

The challenge at this point is to design a molecular system that exhibits these properties. The ideal system will have a rotation rate and hence relaxation time which is sensitive to temperature. In this case a change in sample temperature will change the relaxation time, and ideally a range of spectra similar to the theoretical spectra shown in Figures 3.43 and 3.44 will be observed. Such spectra have been observed in several clathrate compounds of ferrocene (Gibb, 1976). In this section we shall consider some results obtained for $(\eta^6$-cyclohexatriene)($\eta^5$-cyclopentadienyl)-iron(II) hexafluorophosphate and its fluorinated derivative. In these compounds the hexafluorophosphate anions form a static framework within which the cations may undergo rotational processes. The presence of such rotational fluctuations in the electric field gradient is clearly evident in the $^{57}$Fe Mössbauer spectra obtained as a function of temperature as illustrated in Figure 3.45. At 195 K the molecular rotation and hence the fluctuation in the electric field gradient is small and a simple quadrupole split doublet is observed. As the temperature increases the various molecular rotation processes increase and a complex spectrum is observed. As the temperature is further increased to 328 K, the various rotational processes are fast enough to average the electric field gradient to zero and yield a single line spectrum. In the case of the derivative the presence of fluorine on the ring changes the rotational rates and more complex spectra are observed (Fitzsimmons, 1980).

## 9    Conclusions

The use of Mössbauer spectroscopy as a structural probe has changed somewhat over the years. This is to some extent a result of developments in single-crystal X-ray structural determination methods. In the past it was frequently a substantial undertaking to determine the single-crystal structure of a new material, particularly if the material contained many unique atoms and did not have a heavy atom to assist in the analysis. This situation has changed dramatically for three main reasons. Firstly, the development of four-circle single-crystal diffractometers has made the collection of diffraction intensity data almost automatic, secondly, the development of direct methods has made structure solution possible in many cases without the presence of a heavy atom and, thirdly, the advent of high-speed computing facilities has vastly expanded the size of the structure which can be easily and quickly solved.

These developments have greatly influenced the use of Mössbauer spectroscopy as a structural probe. If it is possible to obtain viable single crystals of a new material, it is almost certain that its structure, at least at room temperature, will be known to the Mössbauer spectroscopist. Hence, whereas in the past Mössbauer spectroscopy might have been used to try to deduce the molecular structure of a new material, it is now more likely to be used to study the electronic properties of a material of known structure. Because it is usually much easier to measure the Mössbauer spectrum as a

Fig. 3.45. The Mössbauer spectrum of ($\eta^6$-cyclohexatriene)($\eta^5$-cyclopentadienyl)-iron(II) hexafluorophosphate obtained at various temperatures and illustrating the influence of molecular rotation upon the quadrupole interaction. (From Fitzsimmons & Hume, 1980.)

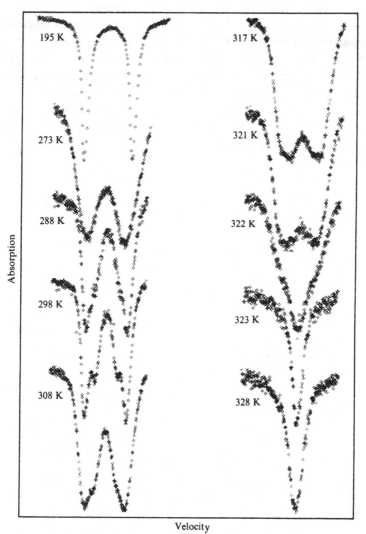

Velocity

function of temperature than to determine single-crystal structure as a function of temperature, Mössbauer spectroscopy is very important in the study of structural phase transformations, especially if these transformations occur at relatively low temperatures, say below room temperature in the case of iron. Several examples have been presented in this chapter which illustrate the sensitivity of Mössbauer spectroscopy to small changes in structure.

In the future it seems quite likely that the combined use of Mössbauer spectroscopy and electron density distributions obtained from X–N calculations will be very valuable in understanding the fine details of the relationship between structure and electronic properties. It seems likely that these developments will rather parallel the complementary use of Mössbauer spectroscopy and magnetic neutron scattering in the study of the details of the magnetic interactions in magnetically ordered systems. In a similar fashion the combined use of Mössbauer spectroscopy, photoelectron spectroscopy, and detailed molecular orbital electronic calculations have proved to be useful in studying the electron density distributions in organoiron clusters, which are important in evaluating chemical interactions at a surface.

# 4

## Mössbauer spectroscopy of magnetic solids

M. F. THOMAS &
C. E. JOHNSON

## 1 Introduction

The aim of this section is to begin with the most general and qualitative picture of the interactions of a Mössbauer nucleus with its environment. By developing and quantifying the treatment of these interactions, we seek to establish connections between the components of the environment and the energies of the nuclear levels and thus to build up an understanding of the features of the environment that shape the Mössbauer spectrum.

### 1.1 Classes of interaction

A schematic picture of the interactions that affect the energy levels of a Mössbauer nucleus is shown in Figure 4.1. In this representation it is seen that the Mössbauer nucleus senses its environment directly through atomic–nuclear interactions [1], solid–nuclear interactions [3] and the external field–nuclear interaction [5]. It is indirectly sensitive to the solid–atom interaction [2] and the external field–atom interaction [4], which are felt through their effect on the nucleus' own atom.

It is usual in magnetic solids for the main influence on the Mössbauer spectrum to be felt through the interactions [1] and [2]. That is, the nucleus senses its own atom and the state of the solid via the effect of the solid on that atom. The interactions of class [1] are (i) the magnetic hyperfine interaction between the atomic spin $S$ and the nuclear spin $I$, (ii) the interaction between the nuclear quadrupole moment $Q$ (which is proportional to the deviation from a spherical distribution of nuclear charge) and the electric field gradient (EFG) produced by the electronic charge distribution of the atom, and (iii) the electrostatic interaction of the nucleus considered as a charged sphere, with the atomic electron density at the nucleus.

Interactions of class [2] affect the number of electrons on the atom and determine the details of the atomic energy levels and electronic wave functions via crystal field and magnetic exchange interactions with neighbouring atoms. As a result of these interactions an atomic spin $S$ is defined. However, further time-dependent interactions in this class, such as spin–lattice and spin–spin relaxation, may cause fluctuations in $S$ such that its effective value $\langle S \rangle$ over the sensing time of the atom–nucleus interaction may cause $\langle S \rangle$ to be reduced below the value $S$ and may even reduce it to $\langle S \rangle = 0$. Interaction [4] between the external magnetic field and the atom plays a role in influencing the atomic state, and in some circumstances can play a major part in influencing the Mössbauer spectrum, for example in the case of a fast relaxing paramagnet, as will be discussed below.

The nucleus senses the atoms of the solid directly via the interaction of its quadrupole moment with the EFG produced by the charge distribution of these atoms in interactions of class [3] and responds directly to an external magnetic field in interaction [5]. These latter interactions are usually not strong and they only play a major role in the shaping of the Mössbauer spectrum in cases where the atomic spin $S = 0$.

This list of interactions divided into schematic classes is not exhaustive, for example in metals conduction electrons contribute interactions in

Fig. 4.1. Scheme of interactions that affect nuclear energy levels of a Mössbauer nucleus.

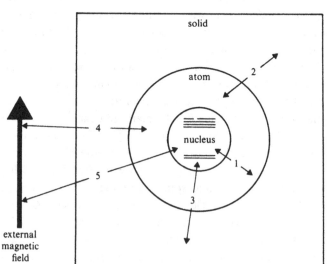

classes [2] and [3], but it does form a basis for the next stage of representing the interactions in a more quantitative form.

### 1.2  The spin Hamiltonian

A framework for treating the nucleus' view of the environment quantitatively and simply has been developed by Abragam & Pryce (1951). The method, called the spin Hamiltonian, utilises the fact that crystal field and spin–orbit interactions often produce a ground state multiplet of atomic levels whose internal splitting is much less than the energy separation between this group of levels and all higher levels. At sufficiently low temperatures, when only the low-lying multiplet is appreciably populated, this group of levels can be considered to be isolated and to contain all the relevant information about the atom. The spin Hamiltonian method defines an effective spin quantum number $S$ within this set of levels such that the number of levels in the multiplet is $2S + 1$. The atomic properties are then described in terms of this effective spin $S$. There is no necessity that this $S$ so defined should be equal to the free atom spin.

The relation of the levels of the spin Hamiltonian to the full set of atomic levels is illustrated in Figure 4.2 for the cases of $Fe^{2+}$ (high spin), $Fe^{2+}$ (low spin) and $Fe^{3+}$ (high spin) ions. These examples reflect the main preoccupation of this chapter with the magnetism of iron-containing compounds. Crystal field interactions, cubic and axial, split the $(3d)^n$ single electron levels as shown in the top section of Figure 4.2. The indicated distribution of the $3d$ electrons in these levels, determined by competition between crystal field and intra-atomic exchange energies determines the nature of the ion and allows energy levels for the whole ion to be drawn. Examples of the lower-lying energy levels for the whole atom are shown in the middle section of Figure 4.2. The lowest lying multiplet of these atomic states, with splittings caused by spin–orbit interaction, is ringed and forms the multiplet of levels considered by the spin Hamiltonian. These states are shown isolated in the bottom section of the diagram, labelled by the spin Hamiltonian spin $S$ and its components $S_z$ along the axis of crystal field symmetry. For each case the spin Hamiltonian levels in zero field are shown on the left and their development in a magnetic field along the symmetry axis is shown on the right.

The energy levels for the states represented by the spin Hamiltonian $\mathscr{H}$ can be written

$$\mathscr{H} = D[S_z^2 - \tfrac{1}{3}S(S+1)] + E[S_x^2 - S_y^2] + \beta \mathbf{B} \cdot \mathbf{g} \cdot \mathbf{s}$$
$$+ \mathbf{I} \cdot \mathbf{A} \cdot \mathbf{S} + \frac{3eQ V_{zz}}{4I(2I-1)}\left[ I_z^2 - \frac{1}{3}I(I+1) + \frac{\eta}{3}(I_x^2 - I_y^2) \right] - g_N \beta_N \mathbf{B} \cdot \mathbf{I}$$

$$(4.1)$$

Where $D$ and $E$ are parameters which describe atomic energy level splitting from axial and non-axial interactions, the subscripts $x$, $y$ and $z$ identify the components of the atomic spin $S$ and the nuclear spin $I$, $\beta$ and $\beta_N$ are the Bohr magneton and the nuclear magneton, $B$ is the magnetic field, $g$ and $g_N$ are the electronic and nuclear $g$-factors, $A$ is the hyperfine coupling tensor, $e$ is the charge on the proton and $V_{zz}$ and $\eta$ are the principal component and the asymmetry parameter of the EFG.

The first three terms in Equation (4.1) describe the ionic multiplet splitting shown in the bottom section of Figure 4.2 in terms of the spin

Fig. 4.2. (*a*) Single 3*d* electron levels for a transition metal atom in a cubic site with axial distortion. (*b*) Energy levels of the (3*d*)$^n$ coupled atomic system. (*c*) Splitting of the lowest atomic level by residual crystal field and spin–orbit interactions, and the applied magnetic field $B$.

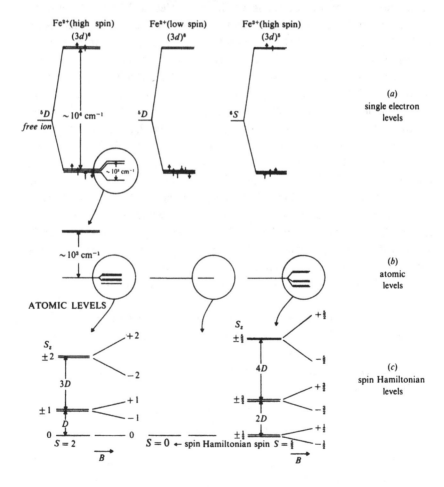

Hamiltonian spin $S$. The final three terms describe the interaction of the nucleus, through its spin $I$ with its environment. In the language of the first section, the first two terms are interactions of class [2]. The third term, representing the interaction between the ion and the magnetic field, is of class [4]. This term is the magnetic splitting of the atomic levels in an applied field in the case of paramagnets and in an exchange field for ordered magnetic materials. Of the terms involving the nucleus, the $I \cdot A \cdot S$ and $-g_N \beta_N B \cdot I$ terms are of class [1] and [5] respectively, while the quadrupole interaction term includes interaction classes [1] and [3], as the electric field gradient at the nucleus has components from the surrounding atom and the neighbouring atoms of the solid.

The last three terms of the spin Hamiltonian shape the Mössbauer spectrum because they describe the interaction of the nucleus with the atom, the solid and the external magnetic field. The term $I \cdot A \cdot S$ describes the magnetic interaction between the nucleus and the atom. This acts through components of the atomic spin which are determined by the Boltzmann population of the spin Hamiltonian states shown in Figure 4.2. Thus the first three terms of the spin Hamiltonian act to determine values for the components of $S$ that are used in the magnetic interaction that shapes the Mössbauer spectrum. The mechanisms involved in this nucleus–atom interaction $I \cdot A \cdot S$ will be discussed in detail in the next section. The quadrupole interaction term represents the interaction of the nuclear quadrupole moment with the EFG produced by the atom and the lattice. The principal component of the EFG is $V_{zz} = \partial^2 V / \partial z^2$ ($V$ is the electric potential at the nucleus) and the asymmetry parameter $\eta = (V_{xx} - V_{yy})/V_{zz}$. For electrons of the nucleus' atom, $V_{zz} = [-e/(4\pi\varepsilon_0)]\langle 3 \cos^2 \theta - 1\rangle\langle r^{-3}\rangle$. The quadrupole interaction is considered in more detail in Chapters 2 and 3. The final term describes the direct interaction of a magnetic field $B$ on the nuclear dipole moment.

Consideration of the spin Hamiltonians in Figure 4.2 shows that in the case of $Fe^{2+}$ (low spin) there will be no magnetic hyperfine interaction $I \cdot A \cdot S$ since $S = 0$. Thus in an applied field $B$, the Mössbauer spectrum will be simply that due to the quadrupole and applied field interactions. In the cases of $Fe^{2+}$ (high spin) and $Fe^{3+}$ (high spin), however, the field $B$ (at a given temperature) acts to determine a value of the atomic spin $\langle S \rangle$ which interacts with the nucleus via the term $I \cdot A \cdot S$. Thus in these cases the spectrum is determined by the effects of the quadrupole interaction and a magnetic interaction that includes both direct and induced components.

## 1.3    *The mechanism of the magnetic hyperfine interaction*

The general form of the magnetic hyperfine interaction used in the spin Hamiltonian, $\mathbf{I} \cdot \mathbf{A} \cdot \mathbf{S}$, links the nuclear and atomic spins by a tensor parameter called the hyperfine coupling tensor, which has elements chosen to give the observed magnetic splittings when combined with the spin Hamiltonian value of $S$. Therefore any inaccuracies introduced by representing the atomic state by the reduced set of spin Hamiltonian levels can be compensated for in the elements of the $A$ tensor. Tensor nature is required for this parameter because the magnetic hyperfine interaction is anisotropic, which means that components of $S$ of the same magnitude but in different directions can produce different magnetic splitting.

Although magnetic hyperfine splitting can be adequately described in terms of the components of the $A$ tensor it is often more convenient to treat the magnetic interactions of the nucleus with both the atom and with an external field $B$, in a similar way, and thus to express the nuclear–atom magnetic interaction in terms of an internal magnetic field called the hyperfine field, $B_{hf}$. For atomic relaxation times short compared with the nuclear sensing time (which is related to $h/A$, where $h$ is the Planck constant; see Chapter 5 for further details) the atomic spin $S$ may be replaced by its average value sensed by the nucleus $\langle S \rangle$. The hyperfine field $B_{hf}$ is defined by the relation

$$g_N \beta_N B_{hf} = -\mathbf{A} \cdot \langle S \rangle \tag{4.2}$$

For a paramagnetic ion in an external field $B$, the total effective field $B_{eff}$ felt by the nucleus is given by

$$\mathbf{B}_{eff} = \mathbf{B} + \mathbf{B}_{hf} = \mathbf{B} - \frac{\mathbf{A} \cdot \langle S \rangle}{g_N \beta_N} = \mathbf{B} + \frac{\langle S \rangle}{S} B_{hf}^0 \tag{4.3}$$

where the saturated hyperfine field $B_{hf}^0$ is equal to $AS/g_N\beta_N$, where $A$ is the magnitude of the principal component of the hyperfine coupling tensor. The interaction mechanisms contained in the hyperfine coupling tensor have been discussed in detail by Marshall and Johnson (Marshall & Johnson, 1962; Johnson, 1971). The three main mechanisms are:

(i) The contact interaction whereby a non-zero spin on a part-filled atomic shell, $3d$ for transition metals and $4f$ for rare earths, polarises the $s$ electrons which have a finite density at the nucleus to give a net spin density at the nuclear site. In such an interaction a contact field $B_c$ can be written

$$\mathbf{B}_c = -2\frac{\mu_0}{4\pi} \beta \langle r^{-3} \rangle \kappa \langle S \rangle \tag{4.4}$$

where $\langle S \rangle$ is the effective value of atomic spin, $\kappa$ is a parameter describing the polarisation of the $s$ electrons and $\langle r^{-3} \rangle$ is the average of $r^{-3}$ over the $3d$ or $4f$ shells. This component of the hyperfine field is isotropic and for transition metals is usually the main component.

(ii) The magnetic field produced at the nucleus by the orbital motion of the electrons of the part-filled shell. This orbital component $B_{orb}$ can be written

$$\mathbf{B}_{orb} = 2\frac{\mu_0}{4\pi}\beta\langle r^{-3}\rangle\langle \mathbf{L}\rangle \qquad (4.5)$$

Where $L$ is the orbital angular momentum quantum number. This component can be large in rare earth ions (except in $Gd^{3+}$ and $Eu^{2+}$ where $L=0$), but in transition metals the orbital angular momentum is usually quenched by crystal field interactions to give $\langle L\rangle=0$. Spin–orbit interactions restore a small amount of orbital angular momentum to give $\langle L\rangle=(g-2)\langle S\rangle$, so that in the case of $3d$ ions the orbital component $B_{orb}$ can be written

$$\mathbf{B}_{orb} = 2\frac{\mu_0}{4\pi}\beta\langle r^{-3}\rangle(g-2)\langle S\rangle \qquad (4.6)$$

(iii) The magnetic hyperfine interaction produced by the dipolar field of the spins of the $3d$ or $4f$ electrons. This component is written

$$\mathbf{B}_{dip} = 2\frac{\mu_0}{4\pi}\beta\langle r^{-3}\rangle\left[\frac{3\mathbf{r}(\mathbf{S}\cdot\mathbf{r})}{r^2}-\mathbf{S}\right]$$

$$= 2\frac{\mu_0}{4\pi}\beta\langle r^{-3}\rangle\langle 3\cos^2\theta-1\rangle\langle \mathbf{S}\rangle \qquad (4.7)$$

The bracket contains the same dependence on the angular distribution of electrons in the part-filled shell as the quadrupole interaction (see Chapter 2, Section 3.1), and therefore

$$\mathbf{B}_{dip} = \frac{2}{ec^2}\beta\langle S\rangle V_{zz} \qquad (4.8)$$

A spherical distribution of charge in $3d$ or $4f$ orbitals simultaneously produces no quadrupole interaction and no dipolar component of the hyperfine field. A second consequence, following from Laplace's law that $V_{xx}+V_{yy}+V_{zz}=0$, is that

$$(B_{dip})_x+(B_{dip})_y+(B_{dip})_z=0 \qquad (4.9)$$

The hyperfine field $B_{hf}$ is thus composed of contact, orbital and dipolar components

$$\mathbf{B}_{hf} = \mathbf{B}_c + \mathbf{B}_{orb} + \mathbf{B}_{dip} \qquad (4.10)$$

and can be written for transition metals as

$$\mathbf{B}_{hf} = 2\frac{\mu_0}{4\pi}\beta\langle r^{-3}\rangle\left\{-\kappa\langle S\rangle+\langle \mathbf{L}\rangle+\left[3\mathbf{r}\frac{(\mathbf{S}\cdot\mathbf{r})}{r^2}-\mathbf{S}\right]\right\}$$

$$= 2\frac{\mu_0}{4\pi}\beta\langle r^{-3}\rangle[-\kappa+(g-2)+\langle 3\cos^2\theta-1\rangle]\langle S\rangle \qquad (4.11)$$

## 1.4     The Mössbauer spectrum

The relative positions of the absorption lines in the Mössbauer spectrum resulting from the nuclear terms of the spin Hamiltonian depend upon the strengths of the magnetic and quadrupole interactions and upon the angle the effective magnetic field makes with the principal axes of the EFG. Some examples of Mössbauer spectra for the nucleus $^{57}$Fe with quadrupole interaction only, magnetic interaction only and combined magnetic and quadrupole interactions are shown in Figure 4.3. Figure 4.3(a) shows the case of an axially symmetric quadrupole interaction ($\eta = 0$) of negative sign ($V_{zz}$ is negative and thus the $m_I = \pm\frac{1}{2}$ nuclear substates lie higher in energy than the $m_I = \pm\frac{3}{2}$ nuclear substates). There is no magnetic interaction and the axis of quantisation $z$ is the principal axis of the EFG. The hyperfine interaction energies, corresponding to the various $m_I$ substates, are shown relative to the centroid of the nuclear states. This case can occur in zero applied field for a diamagnetic ion for which $S = 0$, or for a fast relaxing paramagnet where, over the nuclear sensing time, the averaged value of the atomic spin $\langle S \rangle = 0$, resulting in zero values of $\mathbf{I} \cdot \mathbf{A} \cdot \mathbf{S}$ and $B_{hf}$.

The spectrum in Figure 4.3(b) is shaped by magnetic interactions only, the quadrupole interaction being zero. There are six lines, corresponding to the transitions allowed by the $\Delta m_I = 0, \pm 1$ selection rule appropriate to the magnetic dipole nature of the 14.4 keV nuclear transition in $^{57}$Fe. The effective magnetic field $B_{eff}$, which determines the axis of quantisation, is the vector combination of the applied field $B$ and the hyperfine field $B_{hf}$. For a diamagnetic ion such as $Fe^{2+}$ (low spin) $B_{hf} = 0$ and any magnetic splitting is due to the applied field alone. For an ordered magnetic material or a paramagnet where the atomic relaxation time is long compared with the nuclear sensing time, $\mathbf{B}_{eff} = \mathbf{B}_{hf}$ if there is no applied field, while for paramagnetic and ordered materials in an applied field $\mathbf{B}_{eff} = \mathbf{B}_{hf} + \mathbf{B}$.

In Figure 4.3(c) both magnetic and quadrupole interactions are present, with the effective field $B_{eff}$ acting along the principal axis of an axially symmetric EFG. The quantisation axis is this common direction. In the case illustrated the electric quadrupole interaction is negative and the line positions are seen to be caused by the addition of the interactions shown in (a) and (b). The level energies relative to the centroids of the $I = \frac{3}{2}$ and $I = \frac{1}{2}$ nuclear states come directly from the nuclear terms $\mathscr{H}_N$ of the spin Hamiltonian. For the excited state $I = \frac{3}{2}$

$$\mathscr{H}_N^e = -g_N^e \beta_N B I_z - g_N^e \beta_N B_{hf} I_z + \tfrac{1}{4} e Q V_{zz} [I_z^2 - \tfrac{1}{3} I(I+1)] \tag{4.12}$$

while for the ground state $I = \frac{1}{2}$ and $Q = 0$.

$$\mathscr{H}_N^g = -g_N^g \beta_N B I_z - g_N^g \beta_N B_{hf} I_z \tag{4.13}$$

where $g_N^e$, $g_N^g$ are the nuclear $g$-factors for the excited and ground states magnetic hyperfine interaction only, (c) combined electric quadrupole and magnetic hyperfine interactions where the axis of the effective magnetic field is parallel to the principal axis of the electric field gradient and the magnetic interaction is considerably stronger than the quadrupole interaction, and (d) combined electric quadrupole and magnetic hyperfine interactions where the direction of the effective magnetic field is perpendicular to the principal axis of the electric field gradient and the quadrupole interaction is considerably stronger than the magnetic interaction. In the expressions for the energy level splittings the nuclear excited state g-factor $g_N^e$ is taken as positive.

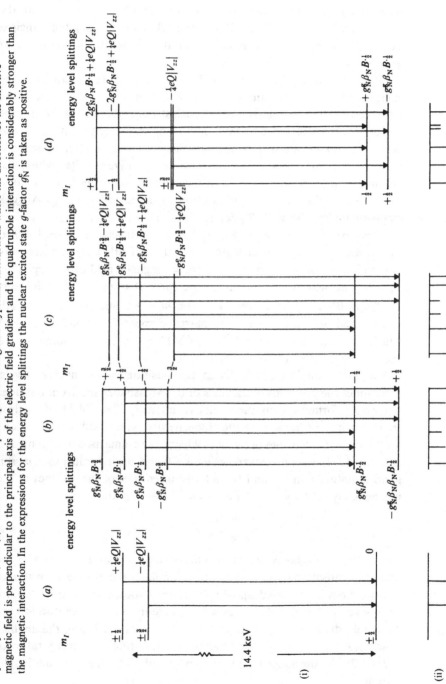

respectively, and the axial symmetry of the EFG gives $\eta = 0$ in the quadrupole interaction. The collinearity of $B_{\text{eff}}$ and $V_{zz}$ along the principal axis of the EFG is a simplifying factor that maintains a good axis of quantisation for the nuclear spin.

In the general case, for $I > \frac{1}{2}$ and where $B_{\text{eff}}$ and $V_{zz}$ are not collinear, the energies of the levels cannot be expressed in simple algebraic form. However, if one of the interactions is much stronger than the other then the energies of the levels can be evaluated using the stronger interaction to define the quantisation axis and treating the weaker interaction as a perturbation. An example of this is shown in Figure 4.3(d) where the dominant interaction is the negative electric quadrupole interaction which defines the $z$ axis of quantisation as the direction of $V_{zz}$. A weak magnetic interaction with $B_{\text{eff}}$ acting along the $x$ axis, perpendicular to $z$, perturbs the nuclear states. In Figure 4.3(d), the nuclear substates, labelled by $m_I$, are not pure, but contain admixtures of other $m_I$ and are only predominantly of the given $m_I$. The weak magnetic field cannot split the $m_I = \pm\frac{3}{2}$ nuclear sublevels since the perturbation matrix elements $\langle \pm\frac{3}{2} | g_N^c \beta_N B_{\text{eff}}^x I_x | \pm\frac{3}{2} \rangle$ are zero. For the $m_I = \pm\frac{1}{2}$ nuclear sublevels admixing and splitting occur via matrix elements $\langle \pm\frac{1}{2} | g_N^c \beta_N B_{\text{eff}}^x I_x | \pm\frac{1}{2} \rangle$ which are equivalent to $\langle \pm\frac{1}{2} | g_N^c \beta_N B_{\text{eff}}^x (I^+ + I^-)/2 | \pm\frac{1}{2} \rangle$ leading to an effective value of $g_N$ of approximately $2g_N^c$.

The spin Hamiltonian, in fixing the position of the nuclear levels, determines the positions of the lines of the Mössbauer spectrum but does not give information on their relative intensity. The 14.4 keV nuclear transition in $^{57}$Fe has a magnetic dipole nature, which leads to expressions for the relative intensities of the lines which are summarised in Greenwood & Gibb (1971). In simple cases, where $\eta = 0$ and $B_{\text{eff}}$ is parallel to $z$, $m_I$ is a good quantum number and the relative intensities of the six lines in the magnetically split spectra of Figures 4.3(b) and (c) are

$$3 : \frac{4 \sin^2 \theta}{1 + \cos^2 \theta} : 1 : 1 : \frac{4 \sin^2 \theta}{1 + \cos^2 \theta} : 3$$

where $\theta$ is the angle between the direction of the gamma-ray beam and $z$, the axis of quantisation of the nuclear spins. Similarly the relative intensities of the two lines in the quadrupole split doublet shown in Figure 4.3(a) are $3 : 1 + 4 \sin^2 \theta / (1 + \cos^2 \theta)$. In a powder, where the axes of quantisation are randomly distributed with respect to the gamma-ray beam, the angular function $4 \sin^2 \theta / (1 + \cos^2 \theta)$ averages to 2, resulting in intensity ratios of $3 : 2 : 1 : 1 : 2 : 3$ for the magnetically split sextet and $1 : 1$ for the quadrupole split doublet.

In cases where non-collinear interactions result in $m_I$ not being a good quantum number, the intensities of the transitions depend upon the degree of $m_I$ mixing in the nuclear sublevels. This leads to the appearance of lines corresponding to the formerly forbidden $\Delta m_I = \pm 2$ transitions as a result of admixed components in previously pure $m_I = \pm \frac{2}{3}$ substates of the excited state. In such cases, the number of transitions in a magnetically split $^{57}$Fe Mössbauer spectrum increases from six to eight.

## 1.5    Examples
### 1.5.1    RbFeF$_4$

As a final step in the consideration of the interactions that shape Mössbauer spectra, we discuss two examples which illustrate some of the ideas introduced above. The first example is the set of spectra of RbFeF$_4$, which contains high spin Fe$^{3+}$, taken over a temperature range from 4.2 to 135 K, that spans the antiferromagnetic ordering (Néel) temperature $T_N = (133 \pm 1)$ K. These spectra are shown in Figure 4.4 and were obtained with a single-crystal absorber, the gamma-ray beam being directed along the crystal $c$ axis. The absence of lines corresponding to the $\Delta m_I = 0$ transitions in the spectra at 4.2 and 78 K show that the quantisation axis of the dominant magnetic interaction is parallel to both the $c$ axis and the gamma-ray beam.

The development of the spectra, as the temperature increases to approach $T_N$, occurs as a result of the change in the ratio of strengths of the magnetic and quadrupole interactions. As the ordering temperature is approached, the value of $\langle S \rangle$ and hence $B_{hf}$ falls steeply, while the quadrupole interaction remains constant and the spectra reflect this change of relative interaction strengths. Analysis of the spectra shows that the gamma-ray beam is parallel to $B_{hf}$, both making an angle of $16.5 \pm 0.5°$ to the principal axis of an axially symmetric EFG. While the magnetic interaction dominates, the $\Delta m_I = 0$ transitions have zero intensity, but when the size of the magnetic and quadrupole interactions become comparable for $120$ K $< T < 133$ K the non-collinearity of $B_{hf}$ and the direction of $V_{zz}$ results in the quantisation axis no longer being parallel to the direction of the gamma-ray beam and the $\Delta m_I = 0$ lines begin to appear in the spectra. At 135 K, above the ordering temperature, the relaxation time of the Fe$^{3+}$ ions is small compared with the nuclear sensing time, giving $\langle S \rangle = 0$ and the disappearance of the magnetic interaction. The gamma-ray beam makes an angle of $16.5°$ with the direction of $V_{zz}$, the axis of quantisation for $T > T_N$, giving a predicted intensity ratio of $3:1.2$ for the lines of the quadrupole split doublet, which agrees well with the observed spectrum. The fitted value of $53.4 \pm 0.5$ T for the low-temperature (saturated) hyperfine field $B_{hf}^0$ reflects the ferric character of the iron atom, as well as the ionic binding of the solid and its two-dimensional magnetic nature, with stronger magnetic exchange interactions within the crystal *ab* planes than between planes in the crystal *c* direction. In the analogous magnetically three-dimensional ferric fluoride $FeF_3$ the saturated hyperfine field is found to be 62.2 T.

Typical values of hyperfine field in $Fe^{3+}$ (high spin) ions in octahedral coordination with increasingly covalent ligands are shown in Table 4.1, while the progression of hyperfine field values in a family of magnetically

Fig. 4.4. Mössbauer spectra of $RbFeF_4$ taken over a range of temperature up to 135 K. The changes in the spectra in the region between 120 and 135 K occur as a result of the changing ratio of the strengths of the electric quadrupole interaction, which is essentially temperature independent, and the magnetic hyperfine field, which decreases quickly as the temperature approaches the Néel temperature of $133 \pm 1$ K. The broadened lines observed in spectra taken just below the Néel temperature are due to changes in hyperfine field within the range of temperature stability of the apparatus. (Rush, Simopoulos, Thomas & Wanklyn, 1976.)

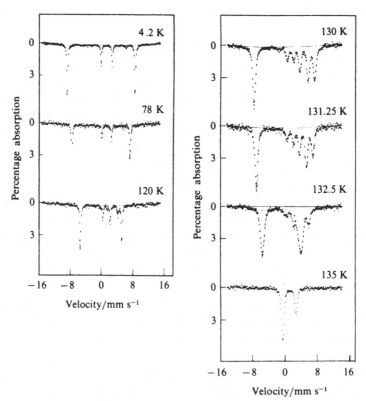

three-, two- and one-dimensional antiferromagnetic ferric fluorides is illustrated in Table 4.2. The effect of increasing covalency is to transfer some of the $3d$ spin to the ligands, thus weakening the polarisation of the iron $s$ electrons and reducing the contact hyperfine field $B_c$. In antiferromagnets any decrease of magnetic dimensionality permits greater freedom of fluctuation for the $Fe^{3+}$ magnet moments which reduces $\langle S \rangle$, with a consequent reduction of the contact hyperfine field. This effect will be discussed in more detail in Section 5.

### 1.5.2 RbFeCl₃

A second contrasting example is of iron as high spin $Fe^{2+}$ in the compound $RbFeCl_3$. The spectra taken at 4.2 and 1.3 K in zero applied field (Figure 4.5) show that magnetic ordering occurs between these temperatures, actually $T_N = 2.55$ K. Both spectra are dominated by the quadrupole interaction, but whereas at 4.2 K the fluctuations of the $Fe^{2+}$ spin over the nuclear sensing time result in $S$, and hence also $B_{hf}$, averaging to zero, at 1.3 K the spectrum shows magnetic splitting and can be fitted with a small hyperfine field of 4.2 T perpendicular to the principal

Table 4.1. *Variation of the saturated hyperfine field with covalency for octahedrally coordinated* $Fe^{3+}$

| Ligand | $B_{hf}^0/T$ |
|---|---|
| $F^-$ | 62.2 |
| $H_2O$ | 58.4 |
| $O^{2-}$ | 54.0 |
| $Cl^-$ | 48.1 |
| $S^{2-}$ | 42.7 |

Table 4.2. *Variation of the saturated hyperfine field with magnetic dimensionality for ferric fluorides*

| Fluoride | Magnetic dimensionality | $B_{hf}^0/T$ |
|---|---|---|
| $FeF_3$ | 3 | 62.2 |
| $RbFeF_4$ | 2 | 53.4 |
| $K_2FeF_5$ | 1 | 41.0 |

axis of the EFG. This results in a splitting of the nuclear sublevels, as illustrated in Figure 4.3(d), which gives rise to the observed spectrum. The hyperfine field of the high spin $Fe^{2+}$ ion can contain contact $B_c$, orbital $B_{orb}$, and dipolar $B_{dip}$ components. The small value of the resultant hyperfine field $B_{hf} = 4.2$ T shows that these components must be of comparable magnitude and partially cancel. In the isomorphous compound $RbFeBr_3$ the cancellation is complete, with $B_{hf} = 0$, and this antiferromagnetically ordered material (as shown by neutron diffraction) gives rise to a

Fig. 4.5. Mössbauer spectra of a single crystal of $RbFeCl_3$ taken with the gamma-ray beam directed perpendicular to the principal axis of the electric field gradient. The spectrum at 4.2 K shows no magnetic interaction. The spectrum at 1.3 K is produced by the situation illustrated in Figure 4.3(d) where a strong electric quadrupole interaction is combined with a weaker magnetic interaction whose effective magnetic field is perpendicular to the principal axis of the electric field gradient. (Baines *et al.*, 1983.)

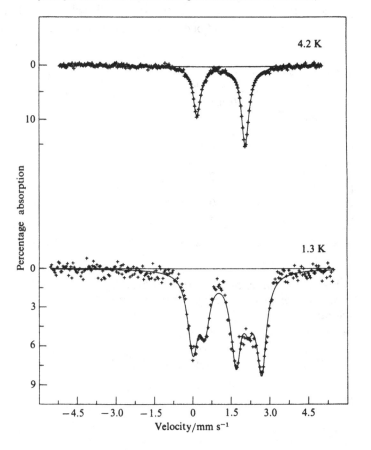

Mössbauer spectrum of a quadrupole split doublet with no indication of any magnetic splitting (Eibschütz, Davidson & Cox, 1973).

### 1.6    *Magnetic features and interactions*

Magnetic phenomena in solids can be regarded as resulting from the interplay of several magnetic features and interactions. The major influences are:

   (i) The state of the magnetic atom, which is determined by the spin Hamiltonian group of atomic levels.

  (ii) The exchange interaction between magnetic atoms, which depends on the separation and relative orientation of their magnetic moments. In its simplest isotropic form the exchange energy between two interacting spins $S_1$ and $S_2$ is given by $-2J\mathbf{S}_1 \cdot \mathbf{S}_2$ where the strength of the coupling is determined by the Coulomb exchange integral $J$. This interaction, summed over interacting neighbours, determines the relative spin directions in the magnetically ordered solid. The exchange interaction can also be represented by an exchange field $B_E$; which acts on individual magnetic moments.

 (iii) The magnetic anisotropy, which determines the orientation of the ordered magnetic spin system relative to the crystal axes. An anisotropy energy constant $K$ can be defined as the energy difference (per atom or per unit volume) between the situations where the atomic spins are directed along the direction in the crystal giving a minimum energy (the magnetic easy axis) and when it is perpendicular to this direction. Alternatively the magnetic anisotropy can be represented in terms of an anisotropy field $B_A$, which is equal to the anisotropy energy constant per atom divided by the atomic magnetic moment.

 (iv) The magnetic dimensionality, which is dependent on any difference in the strength of the exchange interaction along different crystal axes.

These features are not the only influences on magnetic behaviour nor are they independent of each other, but they do supply a framework for discussing the response of the magnetic solid to varying conditions of temperature and applied magnetic field.

In the following sections, examples of the use of Mössbauer spectroscopy in revealing and illustrating different aspects of magnetic behaviour are discussed. These aspects are grouped into sections for ease of reference but many examples contain more than one aspect of magnetic behaviour.

## 2    Studies of magnetic ordering

### 2.1    *Temperature-induced ordering*

Magnetic ordering occurs when the reduction in the magnetic energy per atom on forming an ordered state equals or exceeds the thermal energy per atom. In terms of an atom with spin $S_1$ surrounded by $Z$ equivalent interacting neighbours each with spin $S_2$ an estimate of the ordering temperature $T_{ord}$ (which is the Curie temperature $T_C$ for a ferromagnet and the Néel temperature $T_N$ for an antiferromagnet) is given by

$$-2ZJS_1 \cdot S_2 \simeq k_B T_{ord} \tag{4.14}$$

where $k_B$ is the Boltzmann constant.

A set of Mössbauer spectra spanning the phase transition between the paramagnetic and ordered states is shown for the antiferromagnet $RbFeF_4$ in Figure 4.4. The spectrum at $T = 135$ K, which is above the ordering temperature $T_N = 133 \pm 1$ K, shows quadrupole splitting only, with no magnetic splitting. This occurs since the spin relaxation time of the $Fe^{3+}$ ions is short compared with the nuclear sensing time, which is around $10^{-8}$ s, with the result that the nucleus senses an effective atomic spin $\langle S \rangle = 0$, which results in $B_{hf} = 0$. The quadrupole interaction acts through the electric field gradient from the surrounding ions which is not time dependent.

For the spectra with $T \leqslant 132.5$ K a magnetic hyperfine field $B_{hf}$ exists and is seen to increase as the temperature decreases. This behaviour of $B_{hf}$ follows from the splitting of the spin Hamiltonian states of the $Fe^{3+}$ ion, which can be considered in terms of the exchange field. Thermal population of the split $m_S$ states yields a finite value of $\langle S \rangle$ which increases as the temperature falls, concentrating the population into the lowest lying $m_S = +\frac{3}{2}, +\frac{5}{2}$ substates. Although fast fluctuations in the mean spin value $\langle S \rangle$ still occur, they now do so around a finite value of $\langle S \rangle$ that is sensed by the nuclei as a finite $B_{hf}$, which increases with falling temperature. The exact form of the dependence of $\langle S \rangle$ upon temperature is determined by the details of the crystal field and exchange splitting of the spin Hamiltonian states.

For $Fe^{3+}$ ions the spherical charge distribution results in zero contributions to the magnetic hyperfine field from the orbital component $B_{orb}$ and the dipolar component $B_{dip}$. The hyperfine field is thus solely due to the contact term $B_c$ where

$$B_{hf} = B_c = -2\frac{\mu_0}{4\pi}\beta\langle r^{-3}\rangle\kappa\langle S\rangle \tag{4.15}$$

Thus the variation of $B_{hf}$ with temperature accurately reflects the temperature dependence of $\langle S \rangle$ which corresponds to the sublattice magnetisation. For $Fe^{2+}$ ions where contributions to the hyperfine field may arise from $B_{orb}$ and $B_{dip}$ the hyperfine field may no longer be strictly proportional to the magnetisation.


## 2.2    *Field-induced ordering*

In the materials $RbFeCl_3$ and $CsFeCl_3$ the $Fe^{2+}$ ions form chains that lie along the crystal $c$ axis. The distance between $Fe^{2+}$ ions within the chain is less than half the separation between chains. In $RbFeCl_3$, for temperatures below the Néel temperature of 2.55 K, the $Fe^{2+}$ spins are coupled ferromagnetically (i.e. parallel to each other) within the chains and antiferromagnetically (i.e. antiparallel to each other) between chains. In $CsFeCl_3$ no ordering has been observed down to the lowest temperatures investigated (80 mK).

However, with a magnetic field applied along the chain axis the ordered phase is observed in the Mössbauer spectra at 4.2 K in $RbFeCl_3$ and at 2.2 and 3.0 K in $CsFeCl_3$ (Baines, Johnson & Thomas, 1983). The spectra characteristic of the paramagnetic and ordered states of $RbFeCl_3$ at 4.2 K and 1.3 K in zero applied field are shown in Figure 4.5. These spectra were taken with a single-crystal sample and with the gamma-ray beam perpendicular to the chain axis, which by crystal symmetry is the principal axis of the EFG. From the discussion in Section 1.5.2, it is seen that the 1.3 K spectrum of the ordered phase is characteristic of a weak hyperfine field, which is perpendicular to the axis of the dominant electric quadrupole interaction.

The series of spectra shown in Figure 4.6 indicates the effect of a field $B$ applied along the chain axis of $RbFeCl_3$ at 4.2 K. The spectrum at $B = 0.5$ T shows the paramagnetic response to the field with an induced field acting collinearly with the applied field. The effective field $B_{eff}$ is in a direction parallel to that of $V_{zz}$ and causes greater broadening on the lower energy line characterised by the $m_I = \pm\frac{3}{2}$ substates of the nuclear excited state. At $B = 1$ T the spectrum contains two components. One component has the main splitting on the lower energy line and corresponds to a paramagnetic state. The other component has the main splitting on the higher energy line characterised by the $m_I = \pm\frac{1}{2}$ substates of the nuclear excited state, which is indicative of a weak field perpendicular to $V_{zz}$, and corresponds to the ordered state. This coexistence of paramagnetic and ordered components persists to $B = 3$ T, but the spectra at $B = 4$ T and above show that only the ordered state is present. Similar sets of spectra indicate that in $CsFeCl_3$,

Fig. 4.6. Mössbauer spectra of a single crystal of RbFeCl$_3$ at 4.2 K with an applied field $B$ along the chain axis. The gamma-ray beam is directed perpendicular to this chain axis, which is the principal axis of the electric field gradient. At $B=0.5$ T only the paramagnetic component is observed; for $1.0\,\mathrm{T} \leqslant B \leqslant 3.0\,\mathrm{T}$ there is a coexistence of paramagnetic and ordered components and for $B \geqslant 4.0\,\mathrm{T}$ only ordered conponents are present. (Baines *et al.*, 1983.)

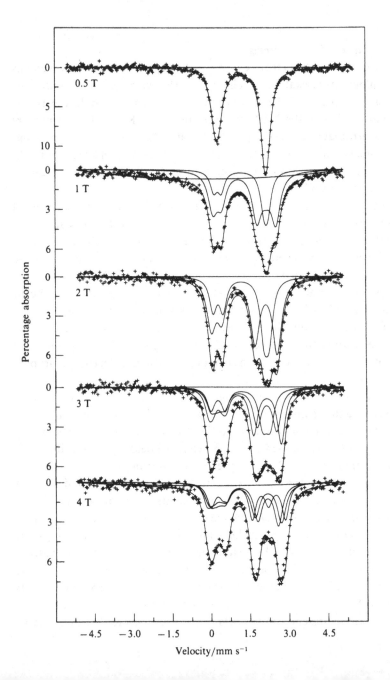

which never orders in zero field, an ordered state is induced by fields along the chain axis for $B \geqslant 3$ T at 2.2 K and $B \geqslant 4$ T at 3.0 K.

The mechanism responsible for this field-induced ordering in $RbFeCl_3$ and $CsFeCl_3$ is the modification of the levels of the $Fe^{2+}$ spin Hamiltonian by the applied field. At the low temperatures of this experiment, only a singlet ground state $(m_S = 0)$ and a doublet $(m_S = \pm 1)$ at an excitation energy $D \sim 10$ K are appreciably populated. The ordering of such a system in both the absence and the presence of an applied field has been discussed by Tsunato & Murao (1971). Magnetic ordering at absolute zero depends upon the relative magnitudes of the excitation energy $D$ and the exchange energy per ion $E_{ex}$. If $D > E_{ex}$ no ordering occurs, while if $D < E_{ex}$ ordering occurs. Thus in $RbFeCl_3$ $E_{ex} > D$ but in $CsFeCl_3$ $E_{ex} < D$. The action of the applied field directed along the chain axis upon the states of the $Fe^{2+}$ ion is illustrated in Figure 4.7 for the case of $CsFeCl_3$. Since the field $B$ is collinear with the quantisation axis it cannot mix the states, but its action is to lower the energy of the $m_S = +1$ state to a level where the non-collinear exchange interaction can mix the $m_S = +1$ and $m_S = 0$ states to form an ordered ground state. A detailed study of the spectra of $RbFeCl_3$ and $CsFeCl_3$ (Baines *et al.*, 1983) shows that the mixing of these states has the

Fig. 4.7. Diagram illustrating the mechanism of field-induced ordering in $CsFeCl_3$. The combined spin–orbit and trigonal crystal field interactions result in a ground state singlet and low-lying excited state doublet characterised as $|0\rangle$ and $|\pm 1\rangle$ respectively. The applied field $B$ reduces the energy of the $|+1\rangle$ state until the exchange interaction energy $E_{ex}$ causes ordering. (Baines *et al.*, 1983.)

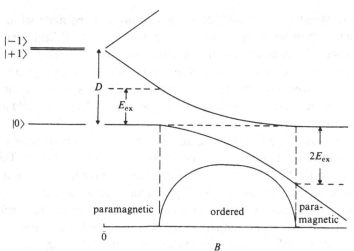

consequence that the quadrupole splitting should decrease with increasing applied field; this is quantitatively confirmed.

An intuitive but fallacious speculation supposes that all paramagnetic materials should become magnetically ordered if taken to a sufficiently low temperature. The most general reason for the failure of this view is that the magnetic interaction energy must be greater than the zero point and crystal field energies to cause ordering, but in special classes of magnetic systems other conditions that prevent magnetic ordering arise. In a one-dimensional magnetic material (see Section 5) the magnetic coupling along the single axis of interaction may be arbitrarily strong without causing long-range ordering. In addition, ordering may not occur in a concentrated iron-containing crystal when the iron atoms increasingly populate the singlet non-magnetic state as the temperature decreases, this corresponds to the case of $CsFeCl_3$ discussed above.

## 3    Magnetic structure

The magnetic structure of an ordered array of spins in a solid is described by the relative alignment of the spins to each other in ferromagnetic, antiferromagnetic, ferrimagnetic or canted arrangements, and the relative orientation of the axis of such a spin system to the crystal axes. The classic technique for studying such magnetic structures is neutron diffraction, but Mössbauer experiments can also give relevant information, both on the relative alignment of the spins and on the directions of the axes of magnetic symmetry within the solid. Four common magnetic structures are shown schematically in Table 4.3, together with a summary of the features which can be used to identify the type of structure from Mössbauer data.

The Mössbauer spectrum of a magnetically ordered material in single-crystal or powder form, below $T_{ord}$ and with no applied field, identifies a ferrimagnetic structure by the superposition of two magnetic six line spectra having different values of hyperfine field $B_{hf}$ and originating from the sublattices with major and minor spin. The spectra of the other structures in these conditions give a single value of hyperfine field and are thus not distinguished.

Ferromagnetic and antiferromagnetic structures can be distinguished by Mössbauer spectra of single-crystal samples in an applied field $B$. This field $B$ adds to the hyperfine field $B_{hf}$ to give the effective field $B_{eff}$, seen in the Mössbauer spectrum as $\mathbf{B}_{eff} = \mathbf{B} + \mathbf{B}_{hf}$ for a ferromagnet and $\mathbf{B}_{eff} = \mathbf{B} \pm \mathbf{B}_{hf}$ for an antiferromagnet. Thus in a ferromagnet a single value of $B_{eff}$ results, but in a structure with an antiferromagnetic axis, the lines of the magnetic spectrum are split into two by the component of $B$ along this axis. The spectra obtained from the canted antiferromagnet $ErFeO_3$ at 4.2 K with zero applied field and with a field of 10 T applied along the antiferromagnetic axis are shown in Figure 4.8.

The distinction between ferromagnetic and antiferromagnetic samples in powder form can still be made by applying a field. The interaction of the applied field $B$ to align a ferromagnetic spin system parallel to the field direction will be stronger than the magnetic anisotropy for all but the smallest values of $B$. Accordingly, in the ferromagnetic case $B_{hf}$ and $B$ are still collinear, to give a single value of $B_{eff}$. In a powder of an antiferromagnet, however, the aligning influence of the applied field (described in Section 4 in relation to the spin reorientation transition) is much weaker and is usually not sufficient to overcome the influence of magnetic anisotropy. Hence, in a powder sample of an antiferromagnet, $B_{hf}$ is randomly oriented with respect to the applied field $B$, giving a range of values of $B_{eff}$, which broadens the lines of the Mössbauer spectrum, distinguishing it from the sharp lines observed with the ferromagnet. The distinction between antiferromagnets and canted antiferromagnets (weak ferromagnets) on the basis of the spin reorientation transition will be discussed in Section 4.

The information contained in the Mössbauer spectrum cannot of itself relate the direction of the magnetic axis to the crystal axes. The spectrum gives information relating the direction of the magnetic axis, via the direction of $B_{hf}$, to the principal axis of the electric field gradient and, for

Table 4.3. *Four common magnetic structures and their identification from magnetically split Mössbauer spectra*

| Ferromagnetic | Antiferromagnetic | Ferrimagnetic | Canted antiferromagnetic (weak ferromagnetic) |
|---|---|---|---|
| ↑ ↑ <br> ↑ ↑ | ↑ ↑ <br> ↓ ↓ | ↑ ↑ <br> ↓ ↓ | ↗↗ <br> ↖↘ |
| No splitting of the spectral lines in an applied field | Splitting of the spectral lines in an applied field directed along the antiferromagnetic axis. Sharp spin reorientation | Distinct magnetically split spectra with different hyperfine fields | Splitting of the spectral lines in an applied field directed along the antiferromagnetic axis. Continuous spin reorientation |

single-crystal samples, to the direction of the gamma-ray beam and any applied magnetic field. Thus a step of linking either the $V_{zz}$ axis, the gamma-ray beam, or the applied field direction, to the crystal axes is needed in order to determine the direction of the magnetic axis within the crystal.

Powder spectra contain information in terms of line positions and intensities, on the relative directions of $B_{hf}$ and $V_{zz}$. This information is summarised in the paper of Kündig (1967) where line positions and intensities are tabulated with respect to the relative strength of the magnetic and electric quadrupole interactions, the angle between the $B_{hf}$ and $V_{zz}$ directions and the asymmetry parameter $\eta$. With the angle between $B_{hf}$ and $V_{zz}$ determined from the spectrum, the relation between the $B_{hf}$ direction and the crystal axes depends upon linking the $V_{zz}$ direction to these axes. Although there is no general rule for this, a knowledge of the crystal

Fig. 4.8. Spectra of a single crystal of ErFeO$_3$ at 4.2 K in (above) zero applied field and (below) with an applied field of 10 T applied along the antiferromagnetic axis. The splitting of the sublattices in the lower spectrum is that expected for the collinear addition and subtraction of the applied field to the hyperfine field. (Prelorendjos, 1980.)

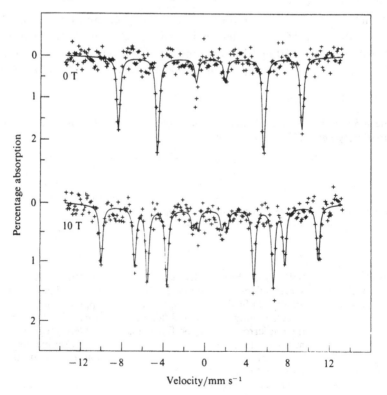

structure may enable the direction of $V_{zz}$ to be deduced; for example, if the magnetic site has an axis of four-fold or three-fold rotational symmetry, this will be the axis of $V_{zz}$.

For single-crystal samples of known orientation, the direction of the Mössbauer gamma-ray beam may be directed along a chosen axis. The relative intensities of the six lines of the magnetically split spectrum are then related to the angle $\theta$ between the magnetic quantisation axis (assuming that the magnetic interaction is dominant) and the direction of the gamma-ray beam as

$$3:\frac{4\sin^2\theta}{1+\cos^2\theta}:1:1:\frac{4\sin^2\theta}{1+\cos^2\theta}:3$$

An example of this technique to establish the orientation of the antiferromagnetic axis in $RbFeF_4$ is shown in Figure 4.4. In the spectra at 4.2 and 77 K the absence of the second and fifth lines indicates that the angle between the magnetic axis and the beam direction is zero and that the antiferromagnetic axis is therefore parallel to the gamma-ray beam, which was directed along the crystal $c$ axis.

In a similar two-step argument, the orientation of the magnetic axis within a crystal can be deduced via its relation to a field $B$ applied along a known crystal axis. The addition of the applied and hyperfine fields of the two sublattices of an antiferromagnet gives rise to two values of the effective field, $B_1$ and $B_2$, which are directly seen in the Mössbauer spectrum. The angle $\alpha$ between the antiferromagnetic axis and the applied field direction is obtained from the relation

$$B_1 - B_2 = 2B_{hf}\cos\alpha \tag{4.16}$$

In an actual experimental geometry with the single crystal in known orientation to both the gamma-ray beam and the applied field, both the splitting and the relative line intensities combine to fix the angular position of the antiferromagnetic axis.

Although the Mössbauer spectra, together with knowledge of the orientation of a sample with respect to the gamma-ray beam and the applied field, can give information on magnetic structure, the method cannot match the detailed information given by neutron diffraction. Thus neutron diffraction can distinguish between different antiferromagnetic configurations or determine the repeating spiral distance of a helical configuration in a way that the Mössbauer method cannot. Justification for using Mössbauer spectroscopy to study magnetic structures lies in its convenience, cheapness and ability to use small single-crystal samples.

## 4     Magnetic phase transitions in antiferromagnets

The configuration of magnetic moments in a solid depends upon external conditions of temperature and applied field. We have discussed above examples of magnetic phase changes between the paramagnetic and the ordered state; in this section we consider in addition magnetic phase changes whereby the spins of an antiferromagnetic system reorient from one crystal axis to another under the influence of an applied field.

Schematic diagrams in Figure 4.9 show the magnetic response (dashed arrows) of an antiferromagnet to a field $B$ directed parallel (*a*) and perpendicular (*b*) to the antiferromagnetic axis. A field collinear with the spins causes an enhancement of the spin value $\langle S \rangle$ of the sublattice parallel to $B$ and a diminuition of $\langle S \rangle$ for the antiparallel sublattice. These changes

Fig. 4.9. Schematic representation of the magnetic response of an antiferromagnet to a field $B$ applied (*a*) parallel and (*b*) perpendicular to the antiferromagnetic axis; $\phi$ is the canting angle of the spins. The behaviour of the magnetic susceptibility $\chi$ as a function of temperature $T$ is shown in (*c*).

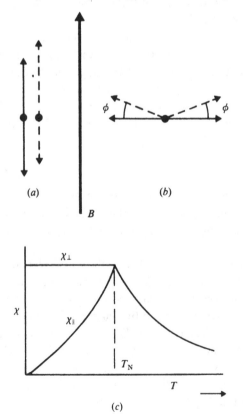

arise from the addition or subtraction of $B$ to or from the exchange field $B_E$ in the argument of the magnetisation function. At low temperatures, the near saturation of the magnetisation function means that the applied field causes little change in $\langle S \rangle$, which is very close to the full spin value $S$. A field perpendicular to the antiferromagnetic axis gives rise to a response by canting the spins. In a collinear antiferromagnet, the canting angle of the spins $\phi$ is given by $\phi = \sin^{-1}[B/(2B_E + B_A)]$, where $B_A$ is the anisotropy field. In contrast to the parallel case, this mechanism suffers no decrease as the temperature is lowered. The response of an antiferromagnet to an applied field and the behaviour of the magnetic susceptibility are shown in Figure 4.9. Below the Néel temperature $T_N$ there are different values of $\chi_{\parallel}$ and $\chi_{\perp}$, the responses with $B$ parallel and perpendicular to the antiferromagnetic axis. At all temperatures below $T_N$ $\chi_{\perp} > \chi_{\parallel}$ and $\chi_{\parallel} \to 0$ as $T \to 0$. The energy consequence of applying a field $B$ to such an antiferromagnetic system is to reduce the energy by an amount $\frac{1}{2}\chi B^2$. Since $\chi_{\perp} > \chi_{\parallel}$ the greater reduction in energy is always produced when $B$ is perpendicular to the coupled spins.

The phenomenon of spin reorientation in antiferromagnets follows directly from the considerations above. In an antiferromagnet the direction of the antiferromagnetic axis with respect to the crystal axis is determined by the magnetic anisotropy energy constant $K$, which is the energy needed to reorient the coupled spins from their equilibrium orientation along the magnetic easy axis, to a perpendicular axis. When a field $B$ is applied along the magnetic easy axis the direction of the coupled spins is determined by whether the energy of the configuration with spins along the easy axis $(-\frac{1}{2}\chi_{\parallel}B^2)$ is greater or less than the energy of the configuration with spins normal to the easy axis $(K - \frac{1}{2}\chi_{\perp}B^2)$.

At low values of applied field, the magnetic anisotropy energy dominates, causing the spins to lie along the magnetic easy axis but, as the applied field is increased, the applied field term causes a greater reduction in energy for the configuration with spins normal to the easy axis. At the spin flop field $B_{sf}$, the energies balance and for $B > B_{sf}$ the lower energy of the configuration with spins normal to the easy axis causes the reorientation of the spins. The spin flop field $B_{sf}$ is related to the exchange field $B_E$ and the anisotropy field $B_A$ by

$$B_{sf} = [2(B_E + B_A)B_A]^{\frac{1}{2}} \tag{4.17}$$

which approximates to $\sqrt{(2B_E B_A)}$, since $B_E \gg B_A$ in almost all magnetic materials.

The geometry of a Mössbauer experiment to detect spin reorientation is depicted in Figure 4.10($a$) and the Mössbauer spectra shown in Figure 4.11 yield information on the angle $\alpha$ of the spins to the applied field, via the relative intensities of the lines and by their splitting, as discussed in Section 3. The variation of the angle $\alpha$ with field is sketched in Figure 4.10(*b*). For a collinear antiferromagnet with the field $B$ applied accurately along the magnetic easy axis, the continuous line shows a sharp spin reorientation (spin flop) at $B = B_{sf}$. For a canted antiferromagnet, the dashed line shows a continuous variation of $\alpha$ with $B$, represented by the relation

$$(2B_A B_E - B^2)\sin\alpha = BB_D \tag{4.18}$$

where the Dzyaloshinski field $B_D$ acts perpendicular to the antiferromagnetic axis, causing the canting of the spins. In the case of the continuous spin reorientation in $ErFeO_3$ at 130 K, illustrated in Figure 4.11, the reorientation curve was fitted to obtain the values: $B_E = 640 \pm 30$ T, $B_A = 0.07 \pm 0.02$ T and $B_D = 10.5 \pm 2.0$ T (Prelorendjos, 1980).

In some magnetic materials the laboratory conditions of temperature

Fig. 4.10. (*a*) The geometry of a Mössbauer experiment to detect spin reorientation. (*b*) The variation with applied field $B$ of the angle $\alpha$ between the applied field and the antiferromagnetic axis for collinear (continuous line) and canted (dashed line) antiferromagnets.

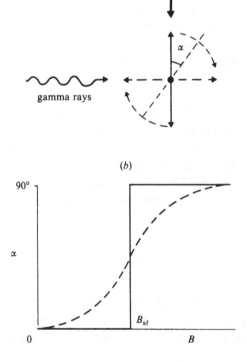

Fig. 4.11. Mössbauer spectra of ErFeO$_3$ taken at 130 K in the geometry of Figure 4.10(a). (Prelorendjos, 1980.) These spectra illustrate the spin reorientation which occurs in a canted antiferromagnet as the applied field increases.

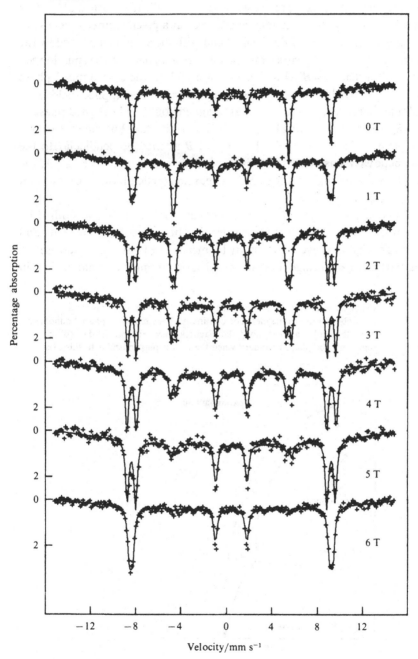

and applied field allow the whole magnetic phase diagram to be investigated. Schematic phase diagrams for a collinear antiferromagnet as a function of temperature and field $B$ applied parallel and perpendicular to the magnetic easy axis are shown in Figure 4.12. It is seen that as $B$ is increased parallel to the magnetic easy axis at a given temperature $T < T_N$ (tracing a vertical line in Figure 4.12 and with the spins as indicated by the arrows) a transition occurs from the antiferromagnetic to the spin flopped phase at a field $B = B_{sf}$. Further increase of $B$ in the spin flopped phase causes increased canting of the spins towards the applied field axis. culminating in a transition to a paramagnetic but highly aligned phase at $B = B_c^{\parallel}$ where the ordering between the spins is broken, but each is strongly coupled to the applied field. For a field $B$ applied perpendicular to the magnetic easy axis there is no spin flopped phase and in increasing field the spins are increasingly canted until the paramagnetic phase is reached at a field $B = B_c^{\perp}$.

The phase transitions that build up such diagrams can be studied by neutron diffraction and by the measurement of magnetic susceptibility and heat capacity. However, their study by means of Mössbauer spectroscopy has particular advantages in that, as well as determining the changes in spin

Fig. 4.12. Applied field $B$ versus temperature $T$ phase diagrams for a Heisenberg antiferromagnet. The continuous lines represent phase boundaries when the field is applied along the magnetic easy axis and the dashed line represents the phase boundary when the field is perpendicular to this axis.

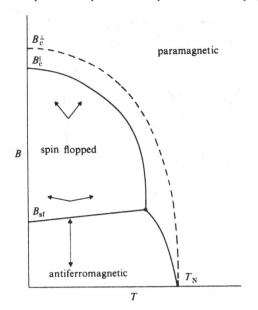

geometry that characterise the phase transitions, the spectra also give information on the excitation energies of the spin configurations.

A study of the magnetic phase transitions induced in a single crystal of the antiferromagnet $Cs_2FeCl_5.H_2O$ as a function of temperature and applied field has been made using Mössbauer spectra to observe the magnetic phases (Johnson, 1985). The sequence of Mössbauer spectra shown in Figures 4.13 and 4.14 were taken at constant temperature with the field applied along the magnetic easy axis; that is, the spectra are illustrating the magnetic conditions along a vertical line in the phase diagram of Figure 4.12. In Figure 4.13 the spectra span the antiferromagnetic to spin flopped transition at 4.2 K. At $B = 1.25$ T the crystal is in the antiferromagnetic configuration with the field acting along the magnetic easy axis, the crystal $a$ axis, causing a splitting of the lines. The gamma-ray beam directed along the crystal $c$ axis is perpendicular to the spins as is seen from the $3:4:1:1:4:3$ relative intensities of the lines. For $B = 1.30$ T the transition is in the process of occurring, as is shown both by the decreased and unresolved splitting of the lines, causing them to appear merely broadened, and by the reduction in

Fig. 4.13. Spectra of a single crystal of $Cs_2FeCl_5.H_2O$ at 4.2 K. The applied field is directed along the magnetic easy axis, the $a$ axis, and the gamma-ray beam is directed along the crystal $c$ axis. At $B = 1.25$ T the antiferromagnetic axis is parallel to the field direction, at $B = 1.30$ T the spin flop is approximately half completed as indicated by the intensities of the $\Delta m_I = 0$ lines and at $B = 1.35$ T the spin flop transition is complete with the antiferromagnetic axis now reoriented along the crystal $c$ axis and parallel to the gamma-ray beam. In this configuration the $\Delta m_I = 0$ lines vanish. (Johnson, 1985.)

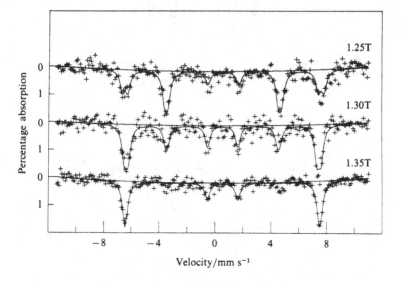

Fig. 4.14. Spectra of a single crystal of $Cs_2FeCl_5 \cdot H_2O$ at 5.4 K. These spectra are taken with the same crystal in the same geometry as for Figure 4.13 but cover the spin flopped to paramagnetic phase transition. At $B=6.0$ T the system is in the spin flopped phase with some canting of the spins toward the field direction producing some intensity in the $\Delta m_I = 0$ lines. For $B \geqslant 7.5$ T the system is in the paramagnetic phase characterised by intensity ratios of 3:4:1:4:3: and with an increasing hyperfine field as the applied field increases. (Johnson, 1985.)

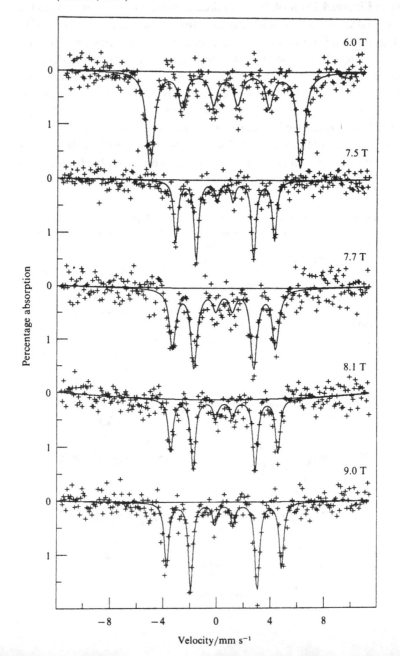

the intensity of the second and fifth ($\Delta m_I = 0$) lines, showing that the spins are reorientating towards the axis of the gamma-ray beam. This reorientation is complete at $B = 1.35$ T, where the zero intensity of the $\Delta m_I = 0$ lines establishes that the antiferromagnetic axis now lies along the axis of the gamma-ray beam, which is the crystal $c$ axis.

The spectra shown in Figure 4.14 illustrate the magnetic configurations at larger fields, with the same experimental geometry and at a temperature of 5.4 K. At the lowest field, $B = 6.0$ T, the material is still in the spin flopped phase, but the finite intensity of the $\Delta m_I = 0$ lines indicates that the spins now cant towards the axis of the applied field and away from the direction of the gamma-ray beam. For $B \geqslant 7.5$ T the system is in the paramagnetic phase. The relative intensities of the lines are 3:4:1:1:4:3, showing that the spins are aligned with the field direction and perpendicular to the gamma-ray beam. The phase transition is marked by a sharp fall in the value of the hyperfine field $B_{hf}$ and in the spectra at $B = 7.7$, 8.1 and 9.0 T the paramagnetic phase shows the spins lying along the applied field axis, with the increasing applied field causing increased values of $\langle S \rangle$ and hence hyperfine field $B_{hf}$.

In the spectra of Figures 4.13 and 4.14, the positions of the lines are determined by the effective field $B_{eff}$ which is related to the applied field $B$ and the hyperfine field $B_{hf}$ by

$$B_{eff}^2 = B_{hf}^2 + B^2 - 2B_{hf}B \sin \alpha \qquad (4.19)$$

where $\alpha$ is the angle of canting of the spins to the applied field, illustrated in Figure 4.9($b$). Fits to the spectra allow $B_{hf}$ and $\alpha$ to be deduced and the complete plot of $B_{hf}$ versus $B$ for $Cs_2FeCl_5 \cdot H_2O$ at $T = 5.4$ K is shown in Figure 4.15, for $B$ both parallel and perpendicular to the magnetic easy axis. The plot illustrates the sharp dips in the value of the hyperfine field at $B = B_{sf}$ and $B = B_c$ the spin flop and paramagnetic transition fields. The values of $B_{sf}$, $B_c^{\parallel}$ and $B_c^{\perp}$ at 0 K, shown in Figure 4.12, are related to the exchange field $B_E$ and anisotropy field $B_A$ of the material through the relations

$$B_{sf} = (2B_E B_A)^{\frac{1}{2}}$$
$$B_c^{\parallel} = 2B_E - B_A \qquad (4.20)$$
$$B_c^{\perp} = 2B_E + B_A$$

In $Cs_2FeCl_5 \cdot H_2O$ values of $B_{sf}$, $B_c^{\parallel}$ and $B_c^{\perp}$ extrapolated to 0 K give values of exchange and anisotropy fields of $B_E \simeq 6.5$ T and $B_A \simeq 0.2$ T respectively. The sharp reduction in $B_{hf}$ at the phase transitions reflects a reduction in $\langle S \rangle$ caused by increased fluctuations of the spins in these conditions. At these transitions, the energy needed to create spin excitations in the spin

Fig. 4.15. Plot of hyperfine field $B_{hf}$ against applied field $B$ for $Cs_2FeCl_5 . H_2O$ at 5.4 K. Dots represent the data with $B$ parallel to the magnetic easy axis and crosses represent the data obtained with $B$ perpendicular to the magnetic easy axis. Applied fields $B_{sf}$, and $B_c^{\parallel}$ and $B_c^{\perp}$ mark the phase transitions to the spin flopped and paramagnetic phases. (Johnson, 1985.)

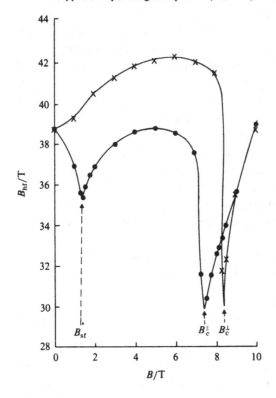

system is reduced and the increased number of quantised spin excitations populated reduces $\langle S \rangle$ and hence $B_{hf}$.

## 5    Low-dimensional magnetism
### 5.1    *Magnetic dimensionality and spin anisotropy*

The general form of the exchange interaction is given by the expression

$$E_{ex} = -\sum_{i,j} 2J_{ij}\mathbf{S}_i \cdot \mathbf{S}_j \qquad (4.21)$$

where the energy is summed over all interacting pairs of atoms, $i$ and $j$, which have spins $S_i$ and $S_j$ and an exchange integral $J_{ij}$. Equation (4.21) can often be appropriately approximated by limiting the summation to nearest neighbours along perpendicular axes. For the case of an orthorhombic magnetic material this leads to

$$E_{ex} \simeq -\sum_i \left( \sum_j 2J^x_{ij} + \sum_j 2J^y_{ij} + \sum_j 2J^z_{ij} \right)(S^x_i S^x_j + S^y_i S^y_j + S^z_i S^z_j) \qquad (4.22)$$

where the summations over $j$ are for nearest neighbours along the $x$, $y$ and $z$ axes and the scalar product of the spins is expanded into components. In Equation (4.22) the first bracket describes the magnetic dimensionality and the second describes the spin anisotropy. In strictly mathematical terms, classes of magnetic dimensionality and spin anisotropy can be defined as in Table 4.4, where the magnetic dimensionality relates to the strength of the exchange coupling along different axes in the crystal and the spin anisotropy relates to the preferred orientations of the spins relative to the crystal axes.

There is no general connection between the magnetic dimensionality and the class of spin anisotropy. In principle the crystal structure and the disposition and type of magnetic atoms allow all possibilities, such as three-dimensional Ising systems or one-dimensional Heisenberg systems. The most usual cause of reduced magnetic dimensionality is that in a particular structure the magnetic atoms are situated close together along an axis (one dimensional) or within a plane (two dimensional) but well separated by non-magnetic atoms from neighbouring chains or planes. The spin anisotropy arises from the multiplicity and separation of the lowest-lying states of the magnetic atoms as well as more complex considerations

Table 4.4. *Classes of (a) magnetic dimensionality and (b) spin anisotropy in ideal magnetic systems*

| (a) | | (b) | |
|---|---|---|---|
| Exchange constants | Magnetic dimensionality | Spin components | Spin anisotropy |
| $J^x = J^y = J^z$ | 3 | $S^x = S^y = S^z$ | Heisenberg (all three spin orientations are equally favourable) |
| $J^x = J^y \neq 0$ $J^z = 0$ | 2 | $S^x = S^y \neq 0$ $S^z = 0$ | X–Y (only spin orientations in a plane are allowed) |
| $J^x = J^y = 0$ $J^z \neq 0$ | 1 | $S^x = S^y = 0$ $S^z \neq 0$ | Ising (only one spin orientation is allowed) |

involving the exchange and dipole–dipole interactions for different spin orientations.

In real solids the exact equalities and the vanishing of particular terms as indicated in Table 4.4 do not occur. However, the classification into corresponding quasi types of magnetic systems, as indicated in Table 4.5, may still be valid in terms of accounting for the experimental magnetic behaviour.

Table 4.5. *Classification of real magnetic materials into quasi systems*

| Exchange constants | Quasi dimensionality | Spin components | Quasi system |
|---|---|---|---|
| $J^x \simeq J^y \simeq J^z$ | 3 | $S^x \simeq S^y \simeq S^z$ | Heisenberg |
| $J^x \simeq J^y \gg J^z$ | 2 | $S^x \simeq S^y \gg S^z$ | X–Y |
| $J^x \simeq J^y \ll J^z$ | 1 | $S^x \simeq S^y \ll S^z$ | Ising |

## 5.2    *Magnetic ordering*

The ability of the above classes of magnetic systems to form a long-range ordered phase at sufficiently low temperature is summarised in Table 4.6. It is seen that a strictly one-dimensional chain of magnetic atoms cannot achieve a state of long-range order, no matter how strong the exchange coupling given by $J^z$, or how large the spin anisotropy. Thus any ordering of a quasi one-dimensional magnetic system occurs as a result of the weaker exchange coupling represented by $J^x$ and $J^y$. In a comprehensive discussion of the magnetism of model compounds (de Jongh & Miedema,

Table 4.6. *The relationship between the magnetic dimensionality, the spin anisotropy and the occurrence of long-range order in ideal magnetic systems*

| Magnetic dimensionality | Spin anisotropy | Long range order |
|---|---|---|
| 1 | Heisenberg | No |
|   | X–Y | No |
|   | Ising | No |
| 2 | Heisenberg | No |
|   | X–Y | No |
|   | Ising | Yes |
| 3 | Heisenberg | Yes |
|   | X–Y | Yes |
|   | Ising | Yes |

1974) it has been shown that a strictly two-dimensional array of Heisenberg magnetic atoms can achieve a state where the magnetic susceptibility goes to infinity without long-range order, but that the introduction of a small degree of spin anisotropy or of a small exchange interaction $J^z$ between layers of magnetic atoms can precipitate ordering.

An example of the information obtainable from a Mössbauer investigation of ordering in a quasi two-dimensional Heisenberg system is the study by Keller & Savic (1983) of $KFeF_4$ and $RbFeF_4$, both of which consist of planes of $(FeF_6)^{3-}$ units separated along the z axis by the alkali metal ions. If the long-range ordering of these systems (detected by the appearance of a magnetically split Mössbauer spectrum) is precipitated by spin anisotropy, the long-range ordering will be in two dimensions, whereas if the ordered state is achieved by means of a small interlayer exchange interaction $J^z$, then the ordering will be in three dimensions. These cases can be distinguished by the shape of the variation of hyperfine field (assumed to be proportional to the sublattice magnetisation) with temperature $T$, as the temperature approaches the ordering temperature $T_N$. The theory of the ordering predicts that in the critical region of temperature, taken in this study as $10^{-2} > (1 - T/T_N) > 5 \times 10^{-4}$, the ratio $B_{hf}/B_{hf}^0$ of the hyperfine field at a temperature $T$ to its saturated value at $0\,K$ is equal to $(1 - T/T_N)^\beta$, where the critical exponent $\beta$ is approximately equal to $\frac{1}{8}$ for two-dimensional ordering and $\frac{1}{3}$ for three-dimensional ordering. The plots of $B_{hf}/B_{hf}^0$ against $T/T_N$ (Figure 4.16) for the two materials show markedly different declining paths as $T$ approaches $T_N$. The analysed values of $\beta = 0.316$ for $RbFeF_4$ and $\beta = 0.151$ for $KFeF_4$ indicate that the ordering is three dimensional in $RbFeF_4$ and two dimensional in $KFeF_4$. This stems from a crystallographic difference in the disposition of the $(FeF_6)^{3-}$ octahedra along the z direction, which leads to a more complete cancellation of the interlayer exchange coupling in the case of $KFeF_4$. Values of $J^z/J^{x,y}$ of approximately $10^{-4}$ for $KFeF_4$ and between $10^{-3}$ and $10^{-2}$ for $RbFeF_4$, have been reported (de Jongh & Miedema, 1974). The investigation of Keller & Savic (1983) is the first study of $KFeF_4$ and $RbFeF_4$ to obtain unambiguous information on the nature of the ordering from the evaluation of the coefficient $\beta$. In obtaining these values, the authors stress the importance of determining $\beta$ from data strictly within the critical region of temperature close to $T_N$. A number of previous investigations obtained inappropriate and inconclusive values of $\beta$ by including data taken at temperatures too far from $T_N$.

### 5.3 *Investigations of quasi one-dimensional magnetic systems*
Quasi one-dimensional magnetic systems with an intrachain

exchange coupling given by $J^z$, which is much greater than the interchain exchange coupling given by $J^{x, y}$, achieve a three-dimensional ordered state as a result of the interchain exchange interaction. Mössbauer studies of such systems investigate the pronounced spin fluctuations which prevent long-range ordering in a pure one-dimensional system and have a large effect on the magnetic behaviour of the ordered phase in a quasi one-dimensional system, particularly on the response of this ordered phase to applied magnetic fields.

A schematic representation of the magnetic structure of the quasi one-dimensional antiferromagnet $K_2FeF_5$ is shown in Figure 4.17. The spins are coupled antiferromagnetically within the chains with exchange constant $J^z$ and between the chains with exchange constant $J^{x, y}$. The three-dimensional ordering temperature occurs at $T_N = 6.95$ K and the ratio of $J^z/J^{x, y}$ has been determined to be approximately $10^3$ (Gupta, Dickson & Johnson, 1978).

Fig. 4.16. Reduced hyperfine field $B_{hf}/B_{hf}^0$ as a function of reduced temperature $T/T_N$. The critical region is expanded in the inset where the solid and dashed lines correspond to fits to $B_{hf}/B_{hf}^0 = (1 - T/T_N)^\beta$ with $\beta = 0.15$ for $KFeF_4$ and $\beta = 0.32$ for $RbFeF_4$. (Keller & Savic, 1983.)

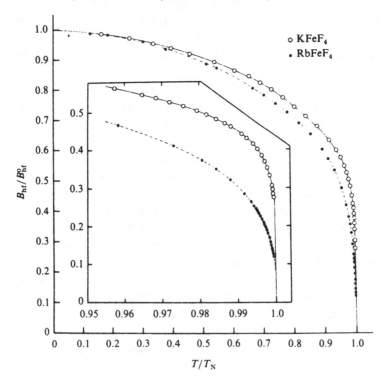

Pronounced spin fluctuations or spin waves occur in the ordered state down to and including absolute zero temperature. These fluctuations reduce the mean spin value $\langle S \rangle$ of the $Fe^{3+}$ ions below $S$. This reduction is directly measurable from the Mössbauer spectra since the magnitude of the hyperfine field $B_{hf}$ is proportional to $\langle S \rangle$. Comparison of the saturated hyperfine field of $41.0 \pm 0.5$ T for $K_2FeF_5$ with that of $62.2 \pm 0.5$ T for $FeF_3$, the corresponding three-dimensional magnetic compound where virtually no spin fluctuations occur, shows that in $K_2FeF_5$ there is considerable reduction of $\langle S \rangle$ below $S$.

The effect of an applied field on the spin fluctuations is exerted largely through the modification of the effective magnetic anisotropy. Magnetic anisotropy is a confining influence on the spins, acting to align them along the magnetic easy axis. Thus the effect of the field in causing a decreased or increased effective magnetic anisotropy will cause enhanced or diminished spin fluctuations. A field applied along the antiferromagnetic axis (normal to the chain of atoms) decreases the effective magnetic anisotropy and the fluctuations increase, thus reducing $\langle S \rangle$ and hence $B_{hf}$. If the field is applied along the chain direction, perpendicular to the magnetic easy axis, the effect is to increase the effective magnetic anisotropy, reducing the fluctuations of the spins in the direction parallel to the field and also canting the spins towards the field direction; these effects are shown in Figure 4.17, and in this case $\langle S \rangle$ and $B_{hf}$ are increased. Experimental results illustrating this behaviour are given in Figures 4.18 and 4.19.

The Mössbauer spectra of Figure 4.18 illustrate every aspect of the magnetic response. The waist in the velocity span of the spectra as the applied field is increased shows the successive decrease and increase in

Fig. 4.17. Schematic magnetic structure of $K_2FeF_5$. The diagram on the right-hand side indicates the damping of fluctuations and the canting of spins along the direction of an applied field.

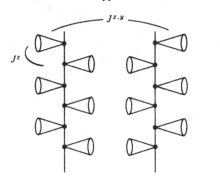
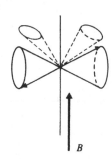

Fig. 4.18. Mössbauer spectra of $K_2FeF_5$ at 4.2 K. The top spectrum is of a powder sample, those below are of a single crystal with the applied field along the magnetic easy axis. (Gupta *et al.*, 1978.)

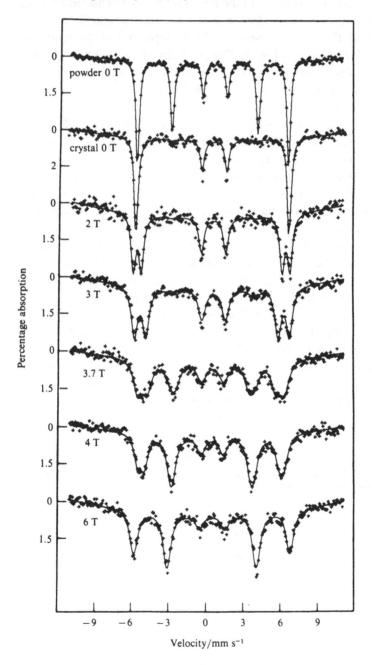

hyperfine field while the spin flop transition is marked by the appearance of the second and fifth ($\Delta m_I = 0$) lines in the spectra for $B > B_{sf}$.

The variation of hyperfine field $B_{hf}$ with field $B$ applied along the magnetic easy axis is shown in Figure 4.19. The magnitude of the hyperfine field is seen to decrease with increasing $B$ up to the spin flop field, $B_{sf} = 3.7$ T, where a spin reorientation occurs. For $B > B_{sf}$ the spins are now perpendicular to the field and further increase of $B$ results in the damping of the fluctuations along the field direction, with the result that $B_{hf}$ increases again.

It is interesting to consider whether the influence of the applied field remains at absolute zero temperature. The experimental result is shown in Figure 4.20, where it is seen that extrapolation to absolute zero indicates that a field parallel to the easy axis has no effect on $\langle S \rangle$ and $B_{hf}$ at 0 K, but that a field perpendicular to the magnetic easy axis continues to decrease the spin fluctuations and increase $\langle S \rangle$ and $B_{hf}$ down to and including 0 K. This behaviour in parallel and perpendicular fields can be explained by considering the susceptibility $\chi$ which represents the response of the system to the applied field. For the field parallel to the antiferromagnetic axis $\chi_\parallel = 0$ at 0 K and the system does not then respond to the applied field, which therefore has no effect. However, for a field applied perpendicular to the

Fig. 4.19. Variation of hyperfine field $B_{hf}$ with applied field $B$ directed along the magnetic easy axis of $K_2FeF_5$. (Gupta *et al.*, 1978.) These data are obtained from the Mössbauer spectra shown in Figure 4.18.

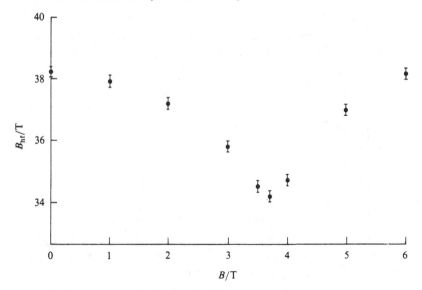

spins $\chi_\perp \neq 0$ at 0 K and the system responds as at finite temperatures and $B_{hf}$ increases with increasing applied field.

The effect of an applied field on the spin fluctuations can also cause a variation of the ordering temperature $T_N$ with applied field. Magnetic ordering occurs when the energy reduction on forming the ordered state equals or exceeds $k_B T$. Following the discussion of Section 2.1, this means that $T_N$ is a function of $\langle S_1 \rangle \cdot \langle S_2 \rangle$, the scalar product of the mean values of the interacting spins. In $K_2FeF_5$ the effect of the applied field depends again upon its direction to the magnetic easy axis. With a field applied parallel to this axis the mean values of the spins $\langle S_1 \rangle$ and $\langle S_2 \rangle$ are reduced as discussed above, thus reducing the exchange energy and causing $T_N$ to fall. With the field applied perpendicular to the magnetic easy axis, two effects occur: (i) $\langle S_1 \rangle$ and $\langle S_2 \rangle$ are increased since the field damps their fluctuations, and (ii) the spins are canted, thus reducing their scalar products $\langle S_1 \rangle \cdot \langle S_2 \rangle$. These effects act in opposite senses on the exchange energy and either may dominate according to the material considered. In the three-dimensional antiferromagnet $FeF_3$ there are virtually no fluctuations to be acted upon and therefore only the canting effect operates and $T_N$ falls with increasing applied field. In $K_2FeF_5$, however, the effect of damping the fluctuations is dominant, the magnitude of the exchange

Fig. 4.20. The effect of an applied field on the magnetisation ($B_{hf}$ against $T$) curve of $K_2FeF_5$. The points □ and △ are for fields applied perpendicular and parallel to the magnetic easy axis respectively, while the points ○ are for zero field. (Boersma *et al.*, 1982.)

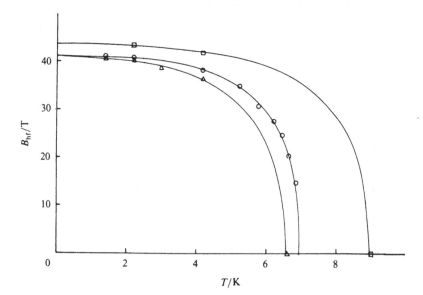

Fig. 4.21. Mössbauer spectra of $K_2FeF_5$ at different temperatures with a field
of 10 T applied along a hard magnetic axis. Ordering is indicated between 8.35
and 8.58 K. (Boersma *et al.*, 1982.)

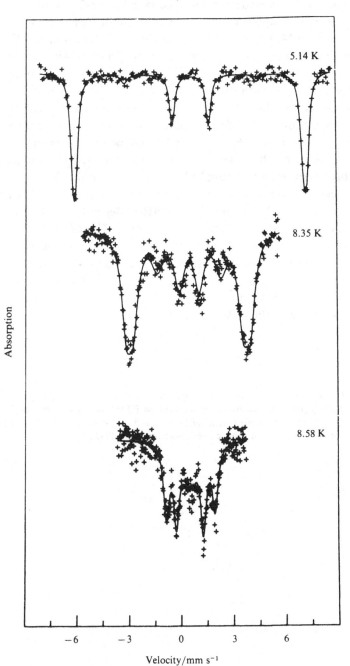

energy is raised with increasing applied field and $T_N$ rises. These effects are shown for $K_2FeF_5$ in Figures 4.20–22. The role of Mössbauer spectroscopy here is to detect the ordered state. The spectra in Figure 4.21 were obtained with a field of 10 T applied perpendicular to the magnetic easy axis at temperatures of 5.14, 8.35 and 8.58 K. Ordering occurs between 8.35 K and 8.58 K, with the magnetic splitting of the 8.58 K spectrum being due solely to the applied field. Figure 4.22 shows the variation of ordering temperature, as a function of applied field, for fields both parallel and perpendicular to the magnetic easy axis. For fields applied parallel to the easy axis a reduction in the ordering temperature is seen as $B$ is increased up to the spin flop value. For $B > B_{sf}$ the spins have become perpendicular to the applied field and the ordering temperature increases with increasing applied field. Increasing fields applied perpendicular to the magnetic easy axis from the start give a steady increase in ordering temperature. The lines shown in Figure 4.22 indicate the calculated dependence of ordering temperature on applied field (Boersma *et al.*, 1982). These calculations use experimentally determined values of $J^z/J^{x, y}$, and anisotropy and exchange fields, but contain no adjustable parameters. The effects of applied fields on the magnetisation curve of $K_2FeF_5$ are summarised in Figure 4.20 which shows the field dependence of both the saturated and unsaturated magnetisation and of the ordering temperature.

Fig. 4.22. The ratio of ordering temperatures with and without an applied field $T_N(B)/T_N(0)$ as a function of the applied field $B$ for $K_2FeF_5$. The points $\triangle$ are for $B$ applied along the easy magnetic axis and the points $\bigcirc$ and $\square$ are for $B$ applied along hard magnetic axes. (Boersma *et al.*, 1982.)

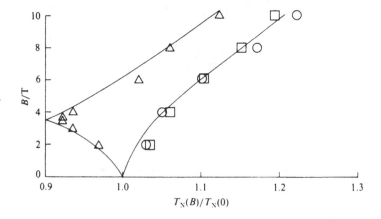

6    **Disordered systems and small particles**
*6.1    Disordered systems*
Disordered magnetic systems fall into three main types:

(i) Crystalline materials exhibiting magnetic order in which some substitution of non-magnetic or different magnetic atoms occurs on sites of the main magnetic species. For small proportions of substitution the general form of the magnetic order remains, but for higher proportions of substitution the magnetic order may be modified or destroyed.

(ii) Spin glasses, which are materials in which magnetic and non-magnetic atoms occupy regular crystalline sites in a random manner. The different distances between magnetic atoms lead to different magnetic interactions between neighbouring magnetic atoms, and at low temperatures a magnetic state is formed embodying both a random character and frustration (which results when adjacent magnetic moments cannot set at optimum orientations with respect to all their neighbours).

(iii) Amorphous magnetic materials, which result when magnetic and non-magnetic atoms aggregate to form a solid with no long-range crystalline order. A random character is introduced into the magnetic interactions by a range of separations between magnetic neighbours and again at low temperatures a magnetic state can result that embodies this random character. In certain amorphous solids and in spin glasses a preponderance of ferromagnetic moments may result in an overall ferromagnetic moment.

A common feature in the $^{57}$Fe Mössbauer spectra of all classes of disordered magnetic materials is the broadening of the lines of the magnetically split sextet caused by a spread in the values of the hyperfine parameters. This spread in values is in turn caused by the range of local environments of the Mössbauer sites and it is normally by analysing these broadenings that the magnetic details of these solids are probed. A comparison of the Mössbauer spectra for crystalline and glassy forms of $Fe_4(P_2O_7)_3$ is shown in Figure 4.23, which illustrates the narrow lines of the crystalline material contrasted with the broadened lines of the glassy solid (G. J. Long, 1984, private communication).

Analytical techniques to extract the form of the hyperfine field distribution from the shape of the broadened lines of the magnetically split spectra of disordered materials were first proposed by Window (1971) and by Hesse & Rubartsch (1974). In the analysis of Window (1971), the hyperfine field distribution, i.e. probability of hyperfine field $p(B_{hf})$ versus hyperfine field $B_{hf}$, is expressed as a series expansion of trigonometric terms, with the fit to the Mössbauer spectrum determining the coefficients of the terms. The technique of Hesse and Rubartsch (1974) is similar, with the hyperfine field distribution being expressed as a smoothed series of step functions. In both analyses, quadrupole interactions are assumed to be sufficiently weak compared with the magnetic hyperfine interaction that they can be treated as a perturbation. An example of spectra exhibiting a distribution of hyperfine fields and fitted by the Window technique is shown in Figure 4.24 together with the associated hyperfine field distributions (Q. A. Pankhurst, 1984, private communication). These spectra are of $FeMgBO_4$, where disorder occurs because of the substitution of about 10% of Mg atoms onto Fe sites and *vice versa*.

While the methods discussed above may provide a good fit to the observed spectra, they typically need to be provided with mean values for the isomer shift and quadrupole splitting and with relative intensities for the outer, middle and inner pairs of lines in the magnetically split sextet spectrum. A different value for these relative intensities will often lead to an equally good fit to the Mössbauer spectrum, with a different, compensating distribution of the hyperfine field. Thus these analyses do not provide a sensitive and reliable method of determining the relative intensities of the

Fig. 4.23. Comparison of Mössbauer spectra for crystalline and glassy forms of $Fe_4(P_2O_7)_3$ at 4.2 K. In the disordered glassy form the lines are much broader than in the crystalline form. A small amount of impurity in the crystalline form is responsible for the weak sextet lying just within the main lines. (G. J. Long, private communication.)

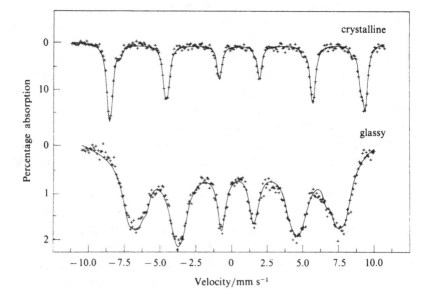

187

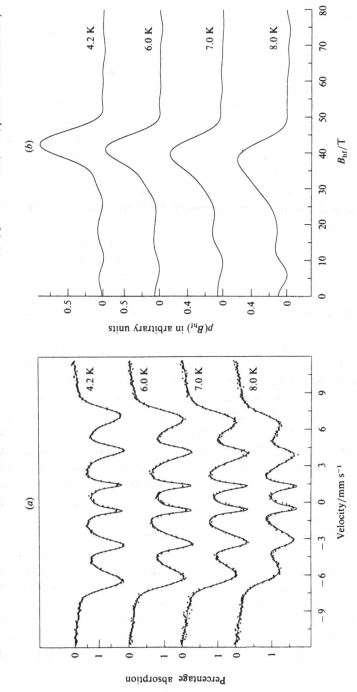

Fig. 4.24. (a) Mössbauer spectra of the disordered compound FeMgBO$_4$ at various temperatures fitted by the Window technique (see text) of hyperfine field distribution. The corresponding hyperfine field distributions are shown in (b). (Q. A. Pankhurst, 1984, private communication.)

lines of the sextet spectrum and hence probing possible preferred spin directions within the solid. The methods are also not sensitive in detecting or verifying correlations between distributions of isomer shift and/or quadrupole splitting with the distribution of hyperfine field. Such correlations could give information on whether distributions of isomer shift, quadrupole splitting and hyperfine field arise from certain repeating types of site in the disordered material, which give characteristic linked ranges of values for the hyperfine parameters, or whether the random character of the sites means that a given value of hyperfine field has no connection with the values of the other hyperfine parameters. These early distribution analyses either assumed no correlations between distributions of hyperfine parameters or imposed fixed correlations and obtained a resulting distribution of hyperfine field which does not necessarily represent a unique solution to fitting the spectrum.

A method involving distributions of hyperfine parameters that probes correlations between them without any *a priori* assumptions has been devised by Lines & Eibschütz (1983). In this method of analysis each line of the magnetic spectrum is separately fitted by a Gaussian distribution of natural-width Lorentzian lines. The line positions give the mean values of the hyperfine parameters, and from asymmetries in the widths of the Gaussian distributions for the outer, middle and inner pairs of lines correlations between distributions of hyperfine field, isomer shift and quadrupole splitting can be extracted, and inferences drawn on the magnetic conditions within the solid. The assumption that the quadrupole interaction can be treated as a perturbation of the dominant magnetic hyperfine interaction is usually made in applying this technique, which has proved to be a subtle and useful tool in probing the conditions within disordered magnetic materials. An important requisite for this analysis to yield reliable results is that the spectra need to be of extremely high quality, having very well defined line positions and widths, and a high degree of instrumental linearity.

Asymmetries in the widths of the outer, middle and inner pairs of lines of a magnetic spectrum caused by a known correlation between distributions of quadrupole and magnetic interactions occurs when a solid contains a range of sites with different asymmetric spatial distributions of surrounding electrons. Such situations can occur in a material containing a range of sites of $Fe^{2+}$ where the $3d$ electron shell assumes different spatial distributions on different sites, or in an alloy such as invar where the nearest-neighbour sites around a given Fe atom can be randomly occupied by Fe or Ni atoms (Window, 1974).

The effect of a single charge and spin dipole moment at a point whose direction from the Mössbauer nucleus makes an angle $\theta$ with the quantisation axis (the direction of the total magnetic hyperfine field) is to give an electric field gradient $V_{zz}$ and dipolar component of the hyperfine field $B_{dip}$, with a similar angular dependence given by

$$V_{zz}(\theta) = (V_{zz}/2)(3\cos^2\theta - 1)$$
$$B_{dip}(\theta) = (B_{dip}/2)(3\cos^2\theta - 1)$$

(4.23)

For a given site, the summation or integration over the whole surrounding asymmetric charge and spin distributions gives total values $V'_{zz}$ and $B'_{dip}$ for the principal component of the EFG and the dipolar hyperfine field respectively. Thus a charge and spin distribution of a given shape gives linked effects of quadrupole and magnetic dipole hyperfine interactions in the Mössbauer spectrum. The results of these correlated effects are demonstrated in Figure 4.25 (after Window, 1974) which shows in successive lines the separate effects on the Mössbauer spectrum of the quadrupole and magnetic dipole hyperfine interactions, then their combined effect at a site and finally the result of summing over different sites. In detail the effect of a negative quadrupole interaction $V'_{zz}$ on the Mössbauer spectrum is shown in (a), and that for a negative $B'_{dip}$ in (b) where the total hyperfine field is increased since the dipolar component has the opposite sign to the dominant contact component. The combined effect of the quadrupole interaction and the dipole hyperfine field is shown in (c) which is obtained by adding (a) and (b). This is the spectrum for a given site. However, different sites have different surrounding electronic charge distributions and for a random ferromagnet like invar the mean values of $V'_{zz}$ and $B'_{dip}$ must be zero. Thus an equally probable site would have values $-V'_{zz}$ and $-B'_{dip}$ and would give the spectrum of line (d). The combined effect of both sites is given in (e) which is obtained by adding (c) and (d) and finally the result of taking into account a range of magnitudes of $V'_{zz}$ and $B'_{dip}$, balanced to give mean values of zero, is shown in (f). This is the predicted result for the correlated distributions and the spectrum exhibits asymmetric widths $\Gamma_i$ of the six lines of the magnetically split sextet such that $\Gamma_1 > \Gamma_6$, $\Gamma_2 < \Gamma_5$ and $\Gamma_3 < \Gamma_4$. Examples of spectra showing such asymmetric widths have been found in a number of studies, as, for example, in the measurements on the invar alloys (Window, 1974; Hesse, Müller & Weichman, 1979).

An ingenious method of studying distributions of quadrupole splitting and isomer shift in the ferromagnetically ordered phase of ferromagnetic amorphous materials, without interference from the distribution of hyperfine fields, has been reported by Kopcewicz, Wagner & Gonser (1984). In this method a radio-frequency field is applied to the sample to reverse the magnetisation, and hence the magnetic hyperfine field at the nucleus, in phase with the applied field. The frequency of this field reversal is greater than the Larmor frequency of the Mössbauer nucleus, with the result that within the sensing time of the magnetic hyperfine interaction the hyperfine field averages to zero and the spectrum reveals the distribution in the remaining quadrupole splitting and isomer shift parameters. The results of this technique applied to the amorphous ferromagnetic glass $Fe_{40}Ni_{40}P_{14}B_6$ is shown in Figure 4.26, where the upper spectrum is that without radio-frequency applied field and the lower spectrum is taken in a field of amplitude 1.3 mT and frequency 44.5 MHz. In the lower spectrum

Fig. 4.25. Schematic Mössbauer spectra to demonstrate how asymmetric line-widths can arise in a magnetic sextet spectrum. The independent effects of the electric quadrupole interaction and the magnetic dipolar contribution to the hyperfine field are shown separately in (*a*) and (*b*) respectively and these effects are summed in (*c*). An equally probable site has a combined electric quadrupole interaction and magnetic dipolar contribution to the hyperfine field of opposite sign and this spectrum is shown in (*d*). Sites with both signs of interaction are present in the material and therefore (*c*) and (*d*) are added to give the spectrum shown in (*e*). Finally the distribution in magnitude as well as sign of the interactions is included in the spectrum shown in (*f*) which models the actual conditions in a real solid. It is seen that the resulting spectrum has linewidths $\Gamma_i$ where $\Gamma_1 > \Gamma_6$, $\Gamma_2 < \Gamma_5$, and $\Gamma_3 < \Gamma_4$.

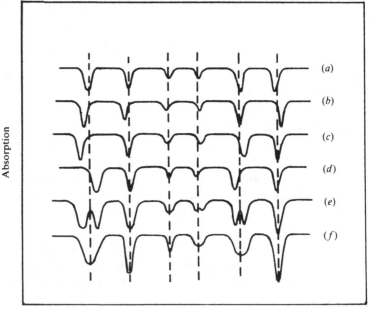

Velocity

the resulting doublet can be analysed for the distribution of quadrupole splitting and isomer shift, and any correlation between these distributions. The sidebands observed in this spectrum are caused by the modulation of the Mössbauer radiation due to magnetostrictive vibrations of the Mössbauer nuclei, with the sidebands occurring at multiples of the applied field frequency. The conditions necessary for the successful use of this method are that the reversing frequency is sufficiently high that complete averaging of the magnetic hyperfine field to zero occurs and that the magnetic materials have a sufficiently low susceptibility that the available fields of up to 2 mT are sufficient to cause reversal of the direction of magnetisation.

### 6.2    Small particles

Small particles or microcrystals of magnetically ordered solids occur widely in ceramics and pigments, and in the technologically important fields of magnetic memory media, catalysts and corrosion products. Mössbauer spectroscopy has proved a most useful tool in

Fig. 4.26. Mössbauer spectra of amorphous $Fe_{40}Ni_{40}P_{14}B_6$ at room temperature: (a) with no radio-frequency field and (b) in a radio-frequency field of frequency 44.5 MHz and amplitude 1.3 mT. In the lower spectrum, the collapse of the hyperfine field and the generation of sidebands is clearly seen. (Kopcewicz *et al.*, 1984.)

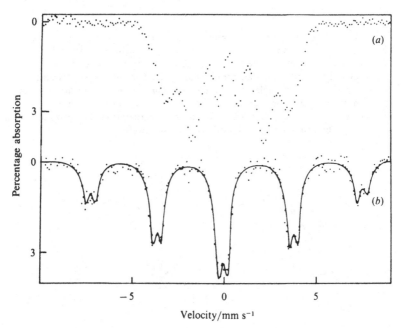

Velocity/mm s$^{-1}$

studying the magnetic behaviour of magnetic solids in this form and the work of Mørup and co-workers (Mørup, Dumesic & Topsøe, 1980) in particular has shown how the main properties of small particle magnetism can be explained on the basis of a model which is presented in simplified form below.

A magnetic anisotropy energy constant $K$ can be identified as the energy difference, per unit volume, between the states where the ordered spins are aligned parallel or perpendicular to the magnetic easy axis. In a given crystal of volume $V$ the anisotropy energy is $KV$. For microcrystals of dimensions $\lesssim 20$ nm the anisotropy energy $KV$ can be comparable with the thermal energy $k_B T$ for $T \lesssim 300$ K. It is the thermally driven fluctuations of the ordered magnetic system within the microcrystal that underlies the magnetic properties of these small particles.

A pictorial representation of the angular fluctuations of the direction of the magnetic spin system about the magnetic easy axis is given in Figure 4.27(a) for a material of uniaxial anisotropy. In this case the energy $E(\theta)$ of displacing this spin direction through an angle $\theta$ from the magnetic easy axis is given by

$$E(\theta) = KV \sin^2 \theta \tag{4.24}$$

which is illustrated in Figure 4.27(b).

The theoretical treatment of the fluctuations can be divided into two regimes. In the first regime, characterised by the condition $k_B T/KV \lesssim 0.1$,

Fig. 4.27. (a) Schematic illustration of the fluctuations of the magnetic axis in a microcrystal of a uniaxial magnetic material. The arrows drawn with continuous lines indicate the fluctuations which give rise to collective excitations and the dashed arrows indicate the fluctuations which lead to superparamagnetic relaxation. The energy $E(\theta)$ associated with the deviation $\theta$ of the magnetic axis from the easy direction is shown in (b), with the representation of collective excitations and superparamagnetic relaxation again being by continuous and dashed arrows respectively.

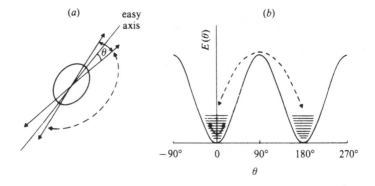

the coupled spins fluctuate through small values of $\theta$ within the well centred on $\theta = 0°$. These fluctuations are represented by the continuous arrows in Figure 4.27. The characteristic time of these fluctuations, known as collective excitations, is much shorter than the Mössbauer sensing time for the magnetic hyperfine interaction and thus the Mössbauer spectrum shows a mean value $\bar{B}_{hf}$ of the fluctuating hyperfine field. The mean hyperfine field acts along the magnetic easy axis and has the value

$$\bar{B}_{hf} = B_{hf}^{max}\langle \cos \theta \rangle \tag{4.25}$$

where $B_{hf}^{max}$ is the full hyperfine field, and the mean value of $\cos \theta$ is given by

$$\langle \cos \theta \rangle = \int_0^{\pi/2} \cos \theta p(\theta) \, d\theta \tag{4.26}$$

where $p(\theta)$ is the normalised probability that the coupled spin direction lies at an angle of $\theta$ to the magnetic easy axis. This $p(\theta)$ is given by

$$p(\theta) = \frac{\exp(-KV \sin^2 \theta/k_B T) \sin \theta}{\int_0^{\pi/2} \exp(-KV \sin^2 \theta/k_B T) \sin \theta \, d\theta} \tag{4.27}$$

Evaluation of the expression for $\langle \cos \theta \rangle$ gives

$$\langle \cos \theta \rangle = 1 - \frac{1}{2}\left(\frac{k_B T}{KV}\right) - \frac{1}{2}\left(\frac{k_B T}{KV}\right)^2 - \frac{5}{4}\left(\frac{k_B T}{KV}\right)^3 - \cdots \tag{4.28}$$

Thus for a particle of a given small volume $V$, $\bar{B}_{hf} \simeq B_{hf}^{max}[1 - (k_B T/2KV)]$ and the effect of the collective excitations is to cause a reduction in the observed hyperfine field. In practice, microcrystalline samples are composed of a distribution of particle volumes which at a given temperature leads to a distribution of hyperfine fields through the relation above. Theoretical spectra exhibiting the spread of hyperfine fields associated with a distribution of microcrystalline volumes are shown in Figure 4.28, where the shape of the asymmetric, inward-curving lines gives information on the distribution of particle volumes.

In the second regime, where typically $k_B T/KV \gtrsim 0.1$, fluctuations of the magnetic spin direction through 180°, as shown by the dashed arrows of Figure 4.27, become predominant, giving the phenomenon of superparamagnetic relaxation. The fluctuation time $\tau$ for these spin reversals is given by

$$\tau \simeq \frac{M\pi^{\frac{1}{2}}}{K\gamma}\left(\frac{k_B T}{KV}\right)^{\frac{1}{2}} \exp\left(\frac{KV}{k_B T}\right) \tag{4.29}$$

where $\gamma$ is the magnetogyric ratio and $M$ is the magnetisation in ferromagnets or the sum of the sublattice magnetisations in antiferromagnets and ferrimagnets. This fluctuation time $\tau$ is typically in the range between $10^{-9}$ and $10^{-11}$ s and is much shorter than the Mössbauer sensing time. Thus in the Mössbauer spectrum the hyperfine field is averaged to zero by these spin reversal fluctuations.

A schematic comparison of the variation of hyperfine field with temperature for a microcrystalline sample and a bulk sample is shown in Figure 4.29. It is seen that, as the temperature rises from zero, the collective excitations cause the observed hyperfine field in the microcrystalline sample to fall below that of the bulk sample. When this reduction is typically 10 to 15% of the bulk hyperfine field, superparamagnetic fluctuations become dominant and the hyperfine field of the microcrystalline sample falls to zero. The temperature at which the hyperfine field falls to zero is called the

Fig. 4.28. Simulated Mössbauer spectra for microcrystalline samples of the uniaxial material α-FeOOH with a broad particle size distribution. The effect of pronounced collective excitations are seen in (*a*) where the mean value of $k_B T/KV = 0.15$; while the effect is much reduced in (*b*) where the mean value of $k_B T/KV = 0.04$.

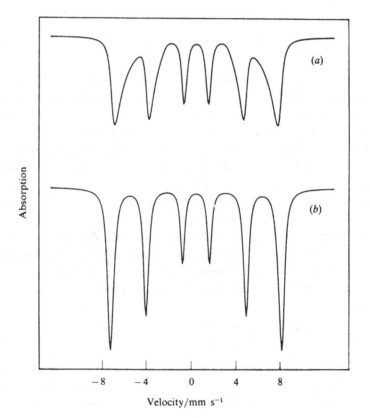

superparamagnetic blocking temperature $T_B$. In the bulk sample the hyperfine field falls to zero at the ordering temperature $T_{ord}$.

Real samples with their distribution of particle volumes often give spectra in which a magnetic sextet component coexists with a component where the hyperfine field has collapsed due to superparamagnetic relaxation. These spectra arise because the superparamagnetic relaxation time $\tau$ is a function of $(KV/k_B T)$. At a given temperature the distribution of microcrystalline volumes within the sample results in most of the particles either having volumes such that the fluctuation time is much greater than the Mössbauer observation time $\tau_{obs}$, giving rise to the magnetically split sextet, or having volumes such that $\tau$ is less than $\tau_{obs}$, giving rise to the component with the collapsed hyperfine field. There are relatively few microcrystals with volumes giving $\tau \sim \tau_{obs}$, and the component from these is not usually visible in the spectrum. As the temperature is increased, the coexistence of the magnetic sextet and collapsed components of the spectrum persist but the intensity of the collapsed component increases at the expense of the magnetic sextet component.

An example of the development of a series of Mössbauer spectra of superparamagnetic samples with increasing temperature is shown in Figure 4.30 (Bell *et al.*, 1984). These spectra are from the biological iron storage materials ferritin and haemosiderin which consist of a microcrystalline iron-containing core enveloped in a protein shell. As the temperature is increased from 4.2 K the spectra first exhibit the asymmetric line shapes caused by the combination of collective excitations with a distribution of particle volumes. At higher temperatures ($\sim 40$ K for ferritin and $\sim 80$ K for haemosiderin) the superparamagnetic fluctuations become important and the spectra are increasingly dominated by the collapsed component, which

Fig. 4.29. Schematic illustration of the variation of hyperfine field $B_{hf}$ with temperature $T$ for microcrystalline and bulk samples.

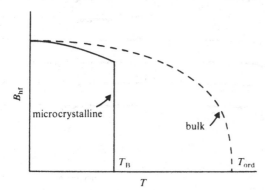

is a quadrupole split doublet. The observation that the blocking temperature for haemosiderin is higher than that for ferritin indicates that the mean value of the microcrystalline anisotropy energy $KV$ is greater in haemosiderin than in ferritin. In order to determine the microcrystalline anisotropy energy per unit volume $K$ (which may differ considerably from that of the bulk material) an independent measurement of the mean particle volume is needed. In this investigation of haemosiderin and ferritin the volumes of the iron-containing cores were measured directly by electron microscopy and revealed that the mean volume of the cores was greater in ferritin than in haemosiderin. The combined results therefore indicate that the microcrystalline anisotropy energy per unit volume $K$ is considerably

Fig. 4.30. Mössbauer spectra of samples of (a) ferritin and (b) haemosiderin, showing the effects of superparamagnetism. Comparison of the two series of spectra shows that the mean superparamagnetic blocking temperature is greater for haemosiderin than for ferritin. (Bell *et al.*, 1984.)

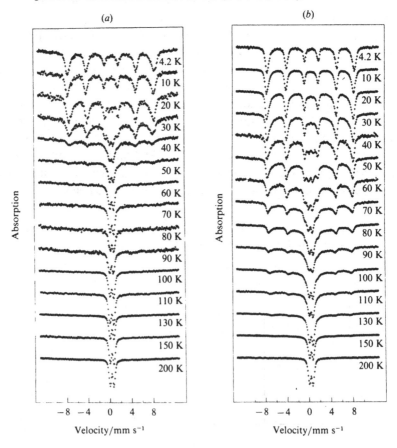

greater for the iron-containing cores of haemosiderin than for those of ferritin.

The Mössbauer spectra of microcrystalline particles of ferromagnetic and ferrimagnetic materials in applied fields can be analysed to yield mean values for the magnetic moment of the particles (Mørup, 1983). The results allow the mean particle volume to be determined if the magnetisation of the material is known.

The effects of interaction between closely packed microcrystals of $\alpha$-FeOOH have been reported by Mørup *et al.* (1983). In this study, the series of Mössbauer spectra obtained with increasing temperature show firstly pronounced asymmetric line shapes and then the collapse of the hyperfine field, without any coexistence of magnetic and collapsed components that characterises the spectra of isolated particles.

## 7    Conclusions and future developments

It has been the aim of this chapter to outline how a Mössbauer nucleus senses its environment and to survey the use of Mössbauer spectroscopy in investigating a range of different aspects of magnetic behaviour in solids.

We think that Mössbauer spectroscopy will be a technique of continuing usefulness in studying magnetic properties of solids and highlight as areas of particular future interest the investigation of disordered magnetic materials, the study of magnetic phase changes induced by temperature or applied magnetic field, the investigation of linear magnetic excitations (magnons) in magnetically ordered substances and the study of non-linear excitations (solitons) in materials of a quasi one-dimensional magnetic character.

# 5

# Time-dependent effects and relaxation in Mössbauer spectroscopy

S. DATTAGUPTA

## 1    Introduction

In this chapter the various kinds of time-dependent phenomena that can be studied by Mössbauer spectroscopy will be reviewed and the relationship between the data obtained by Mössbauer spectroscopy and that derived from other techniques will be assessed.

In many discussions on time-dependent or dynamical effects in Mössbauer spectroscopy the essential physical ideas get obscured by complicated theoretical formalism. The ideas, however, are quite simple and are capable of being appreciated at a qualitative level before embarking on detailed calculations. The present chapter is an attempt to fulfil this need and bring out the important concepts at a non-specialist level. Where necessary, the reader may refer to the original sources cited for more elaborate treatments. For these same reasons, only a few simple models have been chosen, which can be analysed without a detailed mathematical treatment.

As briefly outlined in Chapter 1, the Mössbauer effect concerns the resonant absorption of a gamma ray, of frequency $\omega_0$ and wavevector $k$, by a nucleus. If the absorbing nucleus is located at the position $r$, the electromagnetic field of the gamma ray at the nucleus can be represented by

$$\Phi(r, t) = \text{const} \exp\left[i\mathbf{k} \cdot \mathbf{r} + (i\omega_0 - \Gamma_N)t\right] \tag{5.1}$$

where $\Gamma_N$ is the natural linewidth of the decaying nucleus in its excited state, which is given by $2\pi/\tau_N$, where $\tau_N$ is the mean lifetime of the nuclear excited state. In general, in this chapter, linewidths are expressed in terms of frequency, which must be multiplied by $\hbar$, the Planck constant divided by $2\pi$, in order to convert to an energy. The essential meaning of Equation (5.1) is that the incident gamma ray is not monochromatic at the frequency $\omega_0$ but has a spread $\Gamma_N$, which relates to the uncertainty in the nuclear decay. In other words, this equation can be considered as representing a classical harmonic oscillator corresponding to the electromagnetic wave of the gamma ray, at the frequency $\omega_0$, but damped by an amount $\Gamma_N$.

Now, if the position of the absorbing nucleus only changes as a result of the constant velocity Doppler drive applied to the radioactive source, then

$$\mathbf{k} \cdot \mathbf{r} = -\mathbf{k} \cdot \mathbf{v}t = -\omega t \tag{5.2}$$

where $v$ is the drive velocity and $\omega$ is the Doppler shift in frequency. The total absorption of the gamma ray is obtained by integrating Equation (5.1) over all times, giving

$$\Phi(\omega) = \text{const} \int_0^\infty \exp\left[-\mathrm{i}(\omega - \omega_0)t - (\Gamma_N + \Gamma_I)t\right]\,\mathrm{d}t \tag{5.3}$$

where an additional instrumental linewidth $\Gamma_I$, which incorporates all line-broadening effects associated with the spectrometer, has been added to $\Gamma_N$ to give a total linewidth $\Gamma_T$. The resultant lineshape is Lorentzian, and is given by

$$I(\omega) = \text{Re } \Phi(\omega) = \text{const}\,\frac{\Gamma_T}{(\omega - \omega_0)^2 + \Gamma_T{}^2} \tag{5.4}$$

Equation (5.4) describes a line centred around $\omega_0$ and broadened by an amount $\Gamma_T$ (see Figure 5.1). It may be pertinent to remark that under normal experimental conditions, taking into account both source and absorber linewidths, the total linewidth is given by

$$\Gamma_T \approx 3\Gamma_N \tag{5.5}$$

It is evident from the above discussion that time-dependent considerations assume importance whenever they modulate either (i) the position $r$ of the Mössbauer nucleus over and above the uniform motion

Fig. 5.1. The Mössbauer absorption line given by Equation (5.1).

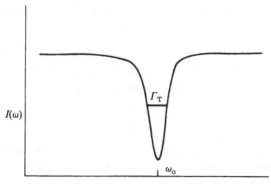

due to the Doppler drive, or (ii) the frequency $\omega_0$ of the Mössbauer radiation. The motion of the Mössbauer nucleus, corresponding to case (i), occurs due to lattice dynamics (i.e. small amplitude motions of the lattice atoms) or diffusional jumps, which affect the position of the atom containing the Mössbauer nucleus. Case (ii) corresponds to relaxation of the surroundings of the Mössbauer nucleus which make the hyperfine interactions fluctuate in time. These cases are examined separately in the subsequent sections, with particular emphasis on relaxation processes. A more detailed consideration of the use of Mössbauer spectroscopy to study the dynamics of nuclei is given in Chapter 6.

## 2    Motion of the Mössbauer nucleus

### 2.1    *Lattice dynamics*

The Mössbauer effect is usually observed in a solid in which the active atom occupies a lattice site. However, such atoms are not stationary but are subject to thermal vibrations around their equilibrium positions which at modest temperatures may be assumed to be harmonic. In view of this motion the phase of the absorbed gamma ray exp $i\mathbf{k} \cdot \mathbf{r}$ becomes a time varying quantity. Although the time average of this quantity over the mean nuclear lifetime would normally be required, $\tau_N$ is found to be significantly larger than the mean time of vibrations $\tau_D$, where $\tau_D \approx 2\pi/\Omega_D$, where $\Omega_D$ is the Debye frequency of the solid. For instance, $\tau_D$ is typically around $10^{-12}$ to $10^{-13}$ s, whereas $\tau_N$ is usually between $10^{-7}$ and $10^{-8}$ s. This means that within the lifetime of the nuclear excited state the Mössbauer atom effectively performs an infinitely large number of vibrations, so that its position becomes completely thermalised. Hence the time average discussed above may be replaced by a thermal average over a stationary probability distribution denoted by the angular brackets below, which leads to an expression for the recoil-free fraction $f$. This is the fraction of emission or absorption events which take place without recoil energy loss; it is also known as the $f$-factor and is related to the Debye–Waller factor.

$$f = \langle \exp i\mathbf{k} \cdot \mathbf{x} \rangle \tag{5.6}$$

where $x$ is the direction of the gamma-ray beam. Assuming for the sake of simplicity that the vibrations are completely isotropic, then

$$f = \left[ \int_{-\infty}^{\infty} \int_{-\infty}^{\infty} P(x, v_x) \exp(ikx) \, dx \, dv_x \right]^3 \tag{5.7}$$

where $P(x, v_x)$ is the stationary probability distribution, which is a function of $x$ and $v_x$, the $x$ components of the position and velocity of the Mössbauer atom. Assuming further that the motion of the atom can be represented by that of a classical harmonic oscillator of mass $m$ and frequency $\Omega$

$$P(x, v_x) = (m\Omega/2k_BT) \exp \left[(-m/2k_BT)(v_x{}^2 + \Omega^2 x^2)\right] \qquad (5.8)$$

where $k_B$ is the Boltzmann constant and $T$ the temperature. Equation (5.7) then yields

$$f = \exp \left[(-\tfrac{3}{2})(k_BT/m\Omega^2)k^2\right] \qquad (5.9)$$

Recalling that for a classical harmonic oscillator

$$\langle r^2 \rangle = \langle (x^2 + y^2 + z^2) \rangle = 3k_BT/m\Omega^2 \qquad (5.10)$$

Equation (5.9) may be expressed as

$$f = \exp \left(-\tfrac{1}{2}k^2 \langle r^2 \rangle \right) \qquad (5.11)$$

Equation (5.11) also holds good for a quantum harmonic oscillator, except that in that case $\langle r^2 \rangle$ has a structure different from that of Equation (5.10). For instance, even at $T = 0$ the zero point motion persists such that

$$\langle r^2 \rangle = 3\hbar/m\Omega \qquad (5.12)$$

Taking the $f$-factor into account in Equation (5.4) the net lineshape is given by

$$I(\omega) = \text{const} \exp \left[(-\tfrac{1}{2}k^2 \langle r^2 \rangle)\Gamma_T\right]/[(\omega - \omega_0)^2 + \Gamma_T^2] \qquad (5.13)$$

In the above analysis it is assumed that the atoms vibrate with a single frequency $\Omega$, as in the Einstein theory, whereas in a real crystal there is a distribution of vibrational frequencies with an appropriate cut-off, as considered in the Debye theory. In any case, it is clear that a measurement of the $f$-factor by Mössbauer spectroscopy can provide knowledge concerning phonon properties, such as their frequency distribution and density of states. Similar information can also be obtained from an analysis of the second-order Doppler shift. Unfortunately, the restriction imposed by the relative timescales, typically $\tau_N \gg \tau_D$ as discussed earlier, normally precludes detailed dynamical examination of the motion of the atoms by Mössbauer spectroscopy alone.

## 2.2 Diffusion

The model of harmonic oscillations of the Mössbauer atom as described above, is quite satisfactory for solids at low temperatures. However, at higher temperatures the small amplitude motion of the atom around its equilibrium site may become unstable such that the atom has a finite probability of making a sudden jump to an adjacent lattice site. Although this type of motion can in principle occur even at very low temperatures via the mechanism of quantum tunnelling, there is little evidence for this phenomenon with Mössbauer isotopes. The diffusive motion can therefore be considered in terms of thermally activated jumps across a barrier of height $E$, so that the average time between successive jumps $\tau_j$ may be described by the classical Arrhenius relation:

$$\tau_j = \tau_0 \exp(E/k_B T) \tag{5.14}$$

where $\tau_0$ is a characteristic time.

There are two distinct mechanisms by which the jumps can occur. In one of these the Mössbauer atom can be considered to occupy an interstitial site, as happens when it is an impurity, with jumps taking place to nearby empty interstitial sites in a random uncorrelated manner (Singwi & Sjölander, 1960). However, a more common possibility is that the atom occupies a lattice site of the host crystal and this can also occur when the Mössbauer atom is an impurity. In these circumstances the atom needs a vacancy, e.g. a Frenkel defect, in order to be able to jump, and the diffusive motion of the active atom becomes highly correlated with that of the vacancies (Bender & Schroeder, 1979, and references therein). Irrespective of whether the jumps occur via the interstitial or the vacancy mechanism the influence on the Mössbauer lineshape is qualitatively the same and involves a random change in the position $r(t)$ of the Mössbauer atom, leading to sudden changes in the phase factor $\exp i\mathbf{k} \cdot \mathbf{r}(t)$. This in turn gives rise to a broadening of the resonance line as discussed below.

In the case of self diffusion of the active atom, such as that of $^{57}$Fe in iron itself, the temperature independent factor $\tau_0$ in Equation (5.14) is of the order of $\tau_D$, the inverse of the Debye frequency, and $\tau_j$ is around $10^{-8}$ s under usual conditions. Since $\tau_j \gg \tau_D$, many oscillations occur before the atom undergoes a jump, and hence it may be considered that between successive jumps of the active atom its oscillatory motion becomes completely thermalised (see Section 2.1). In this situation the average of $\exp i\mathbf{k} \cdot \mathbf{r}(t)$ over the time dependence of $r(t)$ can be factorised into a product of the average over the oscillatory component of $r(t)$, and that over the diffusive component of $r(t)$. The former yields the $f$-factor, as described in Section 2.1, and hence the incident gamma ray can be described by a modified form of Equation (5.1), giving

$$\Phi(k, t) = \text{const } f \exp\left[i(\omega_0 - \Gamma_T)t\right]\langle\exp i\mathbf{k} \cdot \mathbf{r}(t)\rangle \tag{5.15}$$

where the time dependence in $r(t)$ arises solely from the diffusive motion. The meaning of the angular brackets is different from that in Equation (5.1), since they now denote statistical averaging over the random processes which give $r(t)$.

A simple model calculation of the quantity $\langle\exp i\mathbf{k} \cdot \mathbf{r}(t)\rangle$ can now be considered. It is based on random uncorrelated jump diffusion as would apply to interstitial Mössbauer atoms. It has been shown by Chudley &

Elliot (1961) that

$$\langle \exp i\mathbf{k} \cdot \mathbf{r}(t) \rangle = \exp(-\Gamma_{\mathrm{D}} t) \tag{5.16}$$

where $\Gamma_{\mathrm{D}}$, the additional linewidth due to the diffusive motion, is given by

$$\Gamma_{\mathrm{D}} = 2\pi\tau_{\mathrm{j}}^{-1}[1 - P(k)] \tag{5.17}$$

$P(k)$ being the Fourier transform of the probability $P(r)$ of a single jump over a length $r$, where

$$P(k) = \int \exp(i\mathbf{k} \cdot \mathbf{r}) P(r) \, d^3 r \tag{5.18}$$

Substituting Equation (5.16) in Equation (5.15) and integrating over all time (as in Equation (5.3)) also gives a Lorentzian lineshape, represented by

$$\Phi(k, \omega) = \text{const } f/[i(\omega - \omega_0) + (\Gamma_{\mathrm{T}} + \Gamma_{\mathrm{D}})] \tag{5.19}$$

In order to provide a meaningful physical interpretation of the above result $P(k)$ can be calculated in a model in which each jump of the Mössbauer atom is assumed to take it from the centre of a sphere of radius $a$ to any point on its surface. Thus

$$P(r) = (\tfrac{1}{4}a^2)\delta(r - a) \tag{5.20}$$

where $\delta$ is the Dirac delta function. Then, from Equations (5.17) and (5.18),

$$\Gamma_{\mathrm{D}} = 2\pi\tau_{\mathrm{j}}^{-1}[1 - \sin(ka)/(ka)] \tag{5.21}$$

This equation immediately reveals the conditions under which diffusional broadening may be experimentally detectable. It is clear that, firstly, $\tau_{\mathrm{j}}^{-1}$ must be comparable with $\Gamma_{\mathrm{T}}$ or $\Gamma_{\mathrm{N}}$, i.e. the jump time $\tau_{\mathrm{j}}$ must be of the order of the nuclear lifetime $\tau_{\mathrm{N}}$. This is consistent with what would be expected physically in that if $\tau_{\mathrm{j}} \gg \tau_{\mathrm{N}}$ the nucleus would decay before a jump would occur, i.e. the absorbed gamma ray would carry no information about the jump. Secondly, even when the above condition is met, the factor $ka$ cannot be arbitrarily small because then the term inside the square brackets in Equation (5.21) vanishes. Thus it is necessary that $ka > 1$; in the case of $^{57}$Fe, for which $k = 73$ nm$^{-1}$, this is clearly satisfied for values of $a$ around 0.2 to 0.3 nm, which is the sort of distance associated with jumps between nearest-neighbour sites in a solid.

Before closing this section it must be noted that the result of Equation (5.21) must be corrected to account for realistic diffusional effects in a crystal. Firstly, the probability for a single jump $P(r)$ cannot be adequately described by the isotropic picture implied in Equation (5.20), but should be calculated by taking into consideration the crystalline anisotropy (Lewis & Flinn, 1969). Secondly, Equation (5.17) should be modified to take into account the correlated nature of diffusion of the Mössbauer atom via the vacancy mechanism (Bender & Schroeder, 1979).

# 3     Relaxation in Mössbauer spectroscopy

## 3.1     *General considerations*

The dynamical processes described in the previous section take place when the Mössbauer atom physically moves during the lifetime of the Mössbauer nucleus. However, even when the temperature is moderately low in comparison with the melting point of the crystal and diffusive jumps are rather infrequent, there is another source of time-dependent effects which has a strong influence on the spectrum. These arise when the environment of the Mössbauer nucleus changes within its lifetime, thereby altering the frequency $\omega_0$ of the Mössbauer radiation. These processes are usually called relaxation and they occur when the hyperfine interactions, by which the Mössbauer nucleus senses its environment, undergo time-dependent fluctuations.

The hyperfine interactions manifest themselves in three distinct effects: (i) the isomer shift, (ii) the quadrupole splitting, and (iii) the magnetic splitting, and time-dependent fluctuations in any of these three can lead to relaxation phenomena in the Mössbauer spectra. For instance, fluctuations in the electronic charge configuration of the Mössbauer atom can influence both the isomer shift and the quadrupole splitting. These fluctuations can occur due to a variety of causes, such as jump diffusion of a vacancy or an interstitial atom in the vicinity of the Mössbauer atom, dynamic Jahn–Teller distortions, and valence changes. Fluctuations in the magnetic splitting may arise from the dynamical coupling between the electronic spins of the Mössbauer atoms within a system with photons, magnons, or other ionic spins. Such fluctuations may be associated with a wide range of magnetic phenomena. In general, the effects of relaxation on the Mössbauer spectrum can be to produce a shift, broadening or narrowing of the lines, and sometimes even the collapse of the entire spectrum.

It is instructive to consider the predictions of the simplest model of relaxation in which the environment of the nucleus fluctuates in such a manner that the nuclear transition frequency jumps between two values $\omega_1$ and $\omega_2$, each occurring with an equal probability of one-half. Although this model can be most readily thought of in the context of the isomer shift fluctuating between two values, it contains the essential features of relaxation in the case of quadrupole and magnetic splitting as well. If $\omega_R$ is the angular frequency corresponding to the relaxation rate at which the relaxation between $\omega_1$ and $\omega_2$ takes place, the lineshape can be represented by (Dattagupta, 1983)

$$\Phi(\omega) = \mathrm{const}/\{[\Gamma_T + i(\omega - \bar{\omega}) + (\Delta\omega)^2]/[\Gamma_T + i(\omega - \bar{\omega}) + \omega_R]\} \quad (5.22)$$

where $\bar{\omega}$ is the mean frequency:

$$\bar{\omega} = \tfrac{1}{2}(\omega_1 + \omega_2) \tag{5.23}$$

and $\Delta\omega$ is the difference from the mean:

$$\Delta\omega = \tfrac{1}{2}(\omega_1 - \omega_2) \tag{5.24}$$

In deriving Equation (5.22) the effects of diffusion, treated in Section 2.2, have been ignored and the $f$-factor has been absorbed within the constant prefactor. It should be noted that the lineshape is no longer Lorentzian.

The parameter $\omega_R$ characterises the time dependence of the fluctuations in the surroundings of the Mössbauer nucleus. It is therefore dependent on the microscopic interactions of the nucleus with the other degrees of freedom of the entire system of which the nucleus is a part. It is often helpful to consider these effects in terms of a relaxation time $\tau_R$, which is equal to $2\pi/\omega_R$. The nature of the Mössbauer spectrum is drastically affected by the magnitude of $\tau_R$ and the two limiting cases are considered below.

### 3.1.1 The static limit

In this case the relaxation frequency $\omega_R$ is taken to be very small, which is equivalent to the relaxation time $\tau_R$ being very long. This corresponds physically to the situation where the nuclear environment relaxes so slowly that the Mössbauer radiation carries no information about the change in the nuclear environment.

Putting $\omega_R = 0$ in Equation (5.22) gives

$$\Phi(\omega) = \text{const} \left\{ \tfrac{1}{2}/[\Gamma_T + i(\omega - \bar{\omega}) - i\,\Delta\omega] + \tfrac{1}{2}/[\Gamma_T + i(\omega - \bar{\omega}) + i\Delta\omega] \right\}$$
$$\tag{5.25}$$

This equation corresponds to a static spectrum consisting of two Lorentzian lines centred around $(\bar{\omega} + \Delta\omega)$ and $(\bar{\omega} - \Delta\omega)$, which is shown schematically in Figure 5.2. In the context of magnetic splitting, the separation of the spectral lines, quantified by $\Delta\omega$, is related to the Larmor precession frequency and $\tau_L$ the corresponding Larmor precession time, which can be thought of as the measurement time appropriate to the observation of a hyperfine interaction in a Mössbauer spectrum. It is clear that in order to detect the particular interaction it is necessary that $\Delta\omega \gtrsim \Gamma_T$, i.e. $\tau_L < \tau_N$, which is the basic condition to be fulfilled for observing hyperfine structure in Mössbauer spectroscopy. In the case of $^{57}$Fe this implies that $\tau_L \lesssim 10^{-7}$ s.

### 3.1.2 The rapid dynamic limit

In this case the environment relaxes very rapidly, i.e. $\omega_R \gg \Delta\omega$, which automatically implies that $\omega_R \gg \Gamma_T$, in view of the basic condition prescribed above. Equation (5.22) then gives

$$\Phi(\omega) = \text{const}/\{\Gamma_T + i(\omega - \bar{\omega}) + [(\Delta\omega)^2/\omega_R]\} \qquad (5.26)$$

As the quantity $(\Delta\omega)^2/\omega_R$ is real and positive, the spectral line is a single Lorentzian with an additional linewidth $\Gamma_R$, which may be termed the relaxation linewidth, and which is given by $(\Delta\omega)^2/\omega_R$. This situation is shown schematically in Figure 5.3. It should be noted that as the relaxation frequency $\omega_R$ becomes much larger than $(\Delta\omega)^2$, the relaxation linewidth $\Gamma_R$ becomes vanishingly small, leading to motional narrowing of the spectral line.

Therefore, considering the implications of both this and the preceding subsection, as $\omega_R$ increases from a very small value the spectrum first collapses from two lines to a single broad line, which then narrows to the minimum linewidth $\Gamma_T$ as $\omega_R$ becomes very large. The variation in the relaxation rate can be achieved in the laboratory by changing thermodynamic parameters such as the temperature or the pressure.

Fig. 5.2. The Mössbauer absorption lines in the static limit given by Equation (5.25).

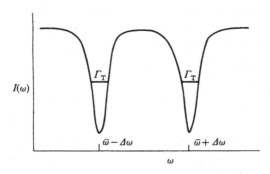

Fig. 5.3. The Mössbauer absorption line in the rapid dynamic limit given by Equation (5.26).

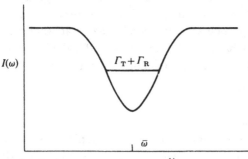

## 3.2    Timescales in relaxation phenomena

The nucleus in a Mössbauer experiment is part of a many-body system consisting of the surrounding electrons and the quasiparticles corresponding to the various other degrees of freedom of the solid. Relaxation effects result from the various time-dependent processes in the vicinity of the nucleus. The nucleus thus acts as a local microscopic probe, which does not participate directly in the relaxation processes in its environment, but which senses these processes via the hyperfine interactions. Now, in interpreting the relaxation behaviour it is necessary to consider the nature and interrelationship of the important timescales of the problem. Some of these timescales are determined by the nature of the Mössbauer isotope and the interaction being studied, i.e., the mean lifetime of the Mössbauer excited state $\tau_N$ and the Larmor precession time $\tau_L$. The other timescales relate to, and are characterised by, the nature of the fluctuations in the nuclear environment. These latter timescales are the inverse of the various relaxation rates and, as mentioned earlier, these can be controlled in the laboratory in various ways. The character of the relaxation spectra obtained obviously depends crucially on the interplay of the various timescales as discussed below.

It has already been noted that in order for resolved magnetic splitting to be observed there must be sufficient time for the nucleus to sense the effects of the magnetic field acting on it. This means that at least one complete Larmor precession must take place before the nucleus decays, i.e. $\tau_L < \tau_N$. Given this restriction it is possible to distinguish between two cases:

(i) $\tau_R \ll \tau_L$; which corresponds to very fast relaxation. Here the hyperfine interaction frequency jumps several times from one value to another before the completion of one Larmor precession. Hence the nucleus senses only a time-averaged hyperfine interaction. This criterion therefore defines the regime in which collapse of the spectra and motional narrowing of the lines takes place. It may be noted that, in view of the restriction that $\tau_L \lesssim \tau_N$ for observation of the hyperfine interaction, the condition $\tau_R \ll \tau_N$ does not necessarily imply that $\tau_R \ll \tau_L$. It is therefore important to emphasise that, contrary to a commonly stated notion, the condition $\tau_R \ll \tau_N$ is not the true criterion for motional narrowing. For instance, although a situation in which $\tau_N \sim 10^{-7}$ s, $\tau_R \sim 10^{-9}$ s and $\tau_L \sim 10^{-10}$ s would obviously satisfy $\tau_L < \tau_N$, as well as the requirements that $\tau_R \ll \tau_N$, it would not satisfy the condition for motional narrowing.

(ii) $\tau_L \ll \tau_R$; which corresponds to very slow relaxation. Here the hyperfine interaction frequency hardly changes its value within one Larmor precession time of the nucleus. The nucleus therefore senses the full hyperfine interaction, thereby yielding a static spectrum. It is important to appreciate that $\tau_L \ll \tau_R$ does not mean that $\tau_R$ is infinitely long, which is of course the extreme static limit. This is illustrated by again considering a situation in which $\tau_N \sim 10^{-7}$ s, $\tau_R \sim 10^{-9}$ s and $\tau_L \sim 10^{-10}$ s, such that a static spectrum ensues, even though the hyperfine interaction frequency may jump a hundred times within the nuclear lifetime. This point also emphasises the fact that a process which appears to be slow in a Mössbauer spectroscopy experiment, may not appear slow when investigated by another technique.

Consider, for instance, a ferromagnetic or antiferromagnetic material studied by Mössbauer spectroscopy, where resolved lines resembling the static spectrum may be observed in the paramagnetic phase above the ordering temperature. This arises because the motion of the electronic spins above the transition point, but not very far from it, may be sufficiently slow to satisfy the condition $\tau_L \ll \tau_R$, resulting in resolved magnetic splitting, while, on a macroscopic timescale, the average spin magnetic moment is zero, i.e. there is no net magnetisation. This argument makes it clear that when using Mössbauer spectroscopy to study magnetic ordering it may be unreliable to identify the ordering temperature as the point at which the magnetic hyperfine spectrum collapses, as this may lead to apparent discrepancies between the values of the ordering temperature estimated by Mössbauer spectroscopy and other methods such as neutron scattering (see Chapter 4).

Finally it is necessary to consider one other point which illustrates a certain limitation of Mössbauer spectroscopy in the investigation of time-dependent hyperfine interactions. Suppose $\tau_N < \tau_R$, which in conjunction with the restriction $\tau_L \lesssim \tau_N$ implies that a static spectrum should be observed. However, the same spectrum would result even if $\tau_R$ becomes very much larger than $\tau_N$. Therefore, beyond a certain limiting value, usually of the order of $\tau_N$, Mössbauer spectroscopy becomes insensitive to changes in $\tau_R$. This implies that very slow changes, i.e. with timescales longer than around $10^{-5}$ s, cannot be effectively studied by Mössbauer spectroscopy.

### 3.3 *More complex timescale considerations*

The classification of the relaxation times becomes somewhat more subtle when a larger set of degrees of freedom have to be taken into account in interpreting the spectra. For instance, consider the magnetic case in which there is an axially symmetric hyperfine coupling between the nuclear and electronic spins, while the effect of the other degrees of freedom of the system on the electronic spins can be described by an effective random field $B_r(t)$. Thus an appropriate spin Hamiltonian for the magnetic hyperfine interaction involving the Mössbauer atom is given by

$$\mathscr{H}(t) = AI_z S_z + g\beta\mathbf{B}_r(t)\cdot\mathbf{S} \tag{5.27}$$

where $A$ is the hyperfine coupling constant, $I_z$ and $S_z$ are the components of the nuclear and electronic spins along the $z$ axis, which is taken to be the symmetry axis of the magnetic hyperfine interaction, $g$ is the electronic $g$-factor and $\beta$ is the Bohr magneton. The various interactions represented by this equation are illustrated schematically in Figure 5.4, which indicates that the origin of $B_r(t)$ may be traced to a coupling between the electronic spin and either (i) other electronic spins, (ii) lattice phonons, (iii) magnons, or (iv) some other degree of freedom of the whole system.

In order to use the Hamiltonian of Equation (5.27) in lineshape calculations it is necessary to specify its spectral properties, which depend, amongst other things, on the correlation function of $B_r(t)$. A simple form of this correlation function is given by

$$\langle B_r^i(0)B_r^j(t)\rangle = \delta_{ij}\langle B_r^2\rangle \exp\left(-t/\tau_C\right) \tag{5.28}$$

where $i$ and $j$ refer to the $x$, $y$ and $z$ axes. Equation (5.28) implies that the different components of $B_r$ are uncorrelated with each other, but that each component is correlated with itself, over a characteristic time $\tau_C$, called the correlation time. The quantity $\langle B_r^2\rangle$, usually referred to as the variance, is a measure of the strength of the effective random field $B_r$. It is clear that $B_r(t)$ produces a random motion in the variable $S_z$, which in turn makes the hyperfine interaction Hamiltonian vary randomly with time. This implies that it might be possible to formulate the problem in terms of an alternative Hamiltonian which is of the form

$$\mathscr{H}'(t) = AI_z S_z(t) \tag{5.29}$$

However, the spectral properties of $S_z(t)$ are quite distinct from those of

Fig. 5.4. A schematic diagram of the various interactions represented by Equation (5.27).

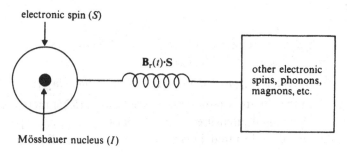

electronic spin ($S$)

$\mathbf{B}_r(t)\cdot\mathbf{S}$

other electronic spins, phonons, magnons, etc.

Mössbauer nucleus ($I$)

$B_r(t)$. The relaxation time $\tau_R$ may in the present case be defined by

$$\tau_R^{-1} = (g^2\beta^2/\hbar^2) \int_0^\infty \langle \mathbf{B}_r(0) \cdot \mathbf{B}_r(t) \rangle \, dt \qquad (5.30)$$

It must be appreciated that $\tau_R$ is very different from the correlation time $\tau_C$; Equations (5.28) and (5.30) give

$$\tau_R^{-1} = (3g^2\beta^2/\hbar^2)\langle B_r^2 \rangle \tau_C \qquad (5.31)$$

Following the above considerations two distinct regimes of relaxation may be discussed:

(i) $\tau_C \ll \tau_L$; in this case the effective random field $B_r(t)$ fluctuates so rapidly that the environment of the Mössbauer atom may be viewed as a source of white noise, with corresponding effects on the magnetic hyperfine interaction. Even within the condition $\tau_C \ll \tau_L$, both slow relaxation, $\tau_R > \tau_L$, or rapid relaxation, $\tau_R < \tau_L$, may occur (Dattagupta, 1983).

(ii) $\tau_C \sim \tau_L$; this situation is more complicated and does not allow a separation of timescales as implied in Equation (5.31). This situation may arise in the spin–spin relaxation in certain rare earth salts (Dattagupta, Shenoy, Dunlap & Asch, 1977).

### 3.4     *Examples of relaxation phenomena in Mössbauer spectra*

In this section examples of observed Mössbauer spectra are presented which illustrate the concepts discussed above. The analysis which will be used requires a slight generalisation of the expression given in Equation (5.22).

The two frequencies $\omega_1$ and $\omega_2$, between which the system relaxes, were originally assumed to occur with the same probability. However, it is more realistic to assign to them probabilities $p_1$ and $p_2$ which may be unequal. In this case the lineshape can be shown to be given by

$$\Phi(\omega) = \text{const}/\{[\Gamma_T + i(\omega - \bar{\omega} - \alpha\Delta\omega)$$
$$+ (\Delta\omega)^2(1-\alpha^2)]/[\Gamma_T + i(\omega - \bar{\omega} + \alpha\,\Delta\omega) + \omega_R]\}$$

$$(5.32)$$

where

$$\alpha = p_1 - p_2 \qquad (5.33)$$

Naturally when $\alpha = 0$ Equation (5.32) reduces to Equation (5.22). The static limit ($\omega_R = 0$) of Equation (5.32) is given by

$$\Phi(\omega) = \text{const}\,\{p_1/[\Gamma_T + i(\omega - \omega_1)] + p_2/[\Gamma_T + i(\omega - \omega_2)]\} \qquad (5.34)$$

In contrast to the situation represented by Equation (5.25) there are now two lines with unequal intensities proportional to $p_1$ and $p_2$ respectively. On the other hand, in the rapid dynamic limit (i.e. $\omega_R \gg \Delta\omega$) Equation (5.32) yields

$$\Phi(\omega) = \text{const}/\{\Gamma_T + i(\omega - \bar{\omega} - \alpha\,\Delta\omega) + [(\Delta\omega)^2(1 - \alpha^2)]/\omega_R\} \quad (5.35)$$

Therefore in the extreme motional narrowing regime (i.e. $\omega_R \to \infty$) the spectrum collapses into a single line which is centred not at $\bar{\omega}$ but at $\bar{\omega} + \alpha\,\Delta\omega$.

Based on the above formulae it is now possible to discuss some experimental cases.

### 3.4.1  Fluctuating isomer shift

We will consider the Mössbauer spectra of CoO and NiO doped with $^{57}$Co, between 300 and 1000 °C (Song & Mullen, 1976). At lower temperatures two lines are observed, which can be identified as being due to the isomer shifts in the charge states $Fe^{2+}$ and $Fe^{3+}$ (Figure 5.5). The low-temperature spectra can be understood on the basis of Equation (5.34) where $\omega_1$ and $\omega_2$ represent the isomer shifts of $Fe^{2+}$ and $Fe^{3+}$ respectively. Since the $Fe^{2+}$ state occurs with a higher probability ($p_1 > p_2$) the corresponding line is more intense. As the source temperature is increased the charge state fluctuates between $Fe^{2+}$ and $Fe^{3+}$ at a rate which can be controlled by varying the partial pressure of oxygen in the furnace. This

Fig. 5.5. Temperature and gas environment dependence of the Mössbauer spectra for a NiO:$^{57}$Co source with a room temperature sodium ferrocyanide absorber. The left-hand set of spectra were obtained in a pure oxygen environment and the sets of spectra to the right were obtained with decreasing partial pressures of oxygen. The dashed lines are approximate representations of the temperature dependence for $Fe^{2+}$ and $Fe^{3+}$ spectra. (Song & Mullen, 1976.)

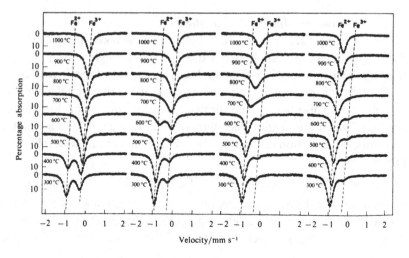

leads to a charge relaxation averaging, under which the two-line pattern merges into a single line with the position of this single line determined by the oxygen partial pressure surrounding the source, in accordance with Equation (5.35). The isomer shift associated with the singlet is intermediate between the $Fe^{2+}$ and $Fe^{3+}$ limits (given by $\omega + \alpha \, \Delta\omega$, *cf.* Equation (5.35)), shifting toward $Fe^{2+}$ for low oxygen partial pressures and towards $Fe^{3+}$ in the high oxygen partial pressure limit (pure oxygen environment). Similar phenomena have been seen in connection with valence averaging in $Eu_3S_4$ (Berkooz, Malamud & Shtrikman, 1968), $EuCu_2Si_2$ (Bauminger *et al.*, 1973) and $Eu_x La_{1-x} Rh_2$ (Bauminger *et al.*, 1974). In these systems electron hopping between a localised $4f$ level and the conduction band results in a Mössbauer resonance line of $^{151}Eu$, intermediate between the expected $Eu^{2+}$ and $Eu^{3+}$ lines (*cf.* Equation (5.35)).

As another example of isomer shift fluctuations we may cite the study of hydrogen diffusion in tantalum metal using the 6.2 keV Mössbauer resonance of $^{181}Ta$ (Heidemann *et al.*, 1976). Here, as the interstitial hydrogen atom jumps into and out of the nearest-neighbour shell of the tantalum nucleus, the isomer shift jumps between two values $\omega_1$ and $\omega_2$. But since the diffusing hydrogen atom has on the average a higher probability of being in a non-nearest-neighbour site the jump rates are unequal. The lineshape can therefore be interpreted on the basis of Equation (5.32). However, the region of experimental interest lies in the rapid dynamic limit and hence the spectrum consists of a single line as described by Equation (5.35) (see Figure 5.6). The relaxation rate $\omega_R$ can in this case be related to the diffusion coefficient of the hydrogen atom in tantalum metal.

### 3.4.2    *Fluctuating electric field gradients*

We consider here the Mössbauer spectra taken between 4 and 550 K for sources of $^{57}Co$ impurity diffused into a single crystal of AgCl (Lindley & Debrunner, 1966). The lineshapes are attributed to a ferrous ion in a substitutional site bound to a positive ion vacancy caused by the removal of a silver ion. This vacancy is constrained to occupy one of the three (1 0 0) positions which are nearest neighbours to the ferrous ion. As a result a quadrupole interaction emerges which can take one of the three following forms:

$$q[3I_x^2 - I(I+1)]; \quad q[3I_y^2 - I(I+1)]; \quad q[3I_z^2 - I(I+1)] \qquad (5.36)$$

where $q$ is proportional to the strength of the interaction and $I_x, I_y$ and $I_z$ are the components of the nuclear angular momentum along the $x$, $y$ and $z$ directions. The energy level structure of $^{57}Fe$ is such that each of the possible forms of the interaction yields a quadrupole split doublet spectrum (*cf.* Figure 1.5 in Chapter 1). The two lines of the doublet are centred around $\pm\Delta\omega$ (see Equation (5.25)) where $\Delta\omega$ corresponds to $3q$ in the present case. This leads to a symmetric doublet in the static limit, in accordance with Equation (5.25), which is what is observed in the experiment between 4 and 230 K (Figure 5.7). With increasing temperature the vacancy starts jumping rapidly between the six equivalent nearest-neighbour positions. This results in the quadrupole interaction changing between the three forms given in Equation (5.36). The resulting lineshape can be interpreted on the basis of

Fig. 5.6. $^{181}$Ta Mössbauer spectra of hydrogen-loaded tantalum foils, with a hydrogen concentration of 1.8%, obtained at different temperatures. (Heidemann *et al.*, 1976.)

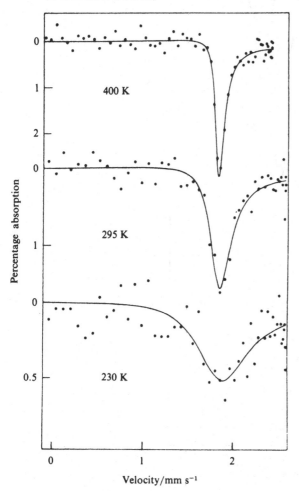

Fig. 5.7. Mössbauer spectra obtained with a source of $^{57}$Co in AgCl at different temperatures. The solid lines in the top three spectra are fits to two Lorentzian lines, with each having a linewidth of 0.6 mm/s. The single line in the 458 K spectrum corresponds to a single Lorentzian line with a linewidth of 0.59 mm/s. (Lindley & Debrunner, 1966.)

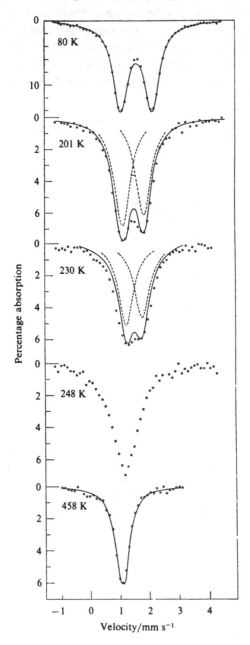

Equation (5.22) where $\omega_R$ can be related to the vacancy jump rate. As $\omega_R$ increases the lines broaden and the spectrum collapses into a single line (see Equation (5.26)). This is seen above 230 K (Figure 5.7). Finally when $\omega_R$ becomes very large the quadrupole interaction averages out to the sum of the three forms given in Equation (5.36), i.e. zero, and the line gets motionally narrowed to the natural width; in agreement with Equation (5.26).

### 3.4.3 Fluctuating magnetic hyperfine fields

We turn our attention in this subsection to an experimental example dealing with relaxation of magnetic hyperfine fields. The example chosen involves the 84.6 keV transition in $^{170}$Yb, in the compound $Cs_2NaYbCl_6$ (Shenoy, Dunlap, Dattagupta & Asch, 1976). The $Yb^{3+}$ ion is in a site of octahedral symmetry such that the electronic ground state is a well-isolated $\Gamma_6$ Kramers' doublet. In the absence of an external magnetic field the hyperfine Hamiltonian can be written as

$$\mathcal{H}_0 = A\mathbf{I}\cdot\mathbf{S} \tag{5.37}$$

where $S = \frac{1}{2}$, and the nuclear spin $I$ for $^{170}$Yb is zero in the excited state and two in the ground state. Equation (5.37) therefore leads to two hyperfine frequencies, $A$ and $3A/2$, which may be identified with $\omega_1$ and $\omega_2$ in the model described by Equation (5.32). The static spectrum is thus given by Equation (5.34) where the probabilities $p_1$ and $p_2$ are $\frac{2}{5}$ and $\frac{3}{5}$ respectively.

We consider now the relaxation effects which occur due to dipolar interactions between $Yb^{3+}$ ions. As a result the central Mössbauer ion may be assumed to be under the influence of an effective random field $B_r(t)$ and the full Hamiltonian may then be written as

$$\mathcal{H}(t) = A\mathbf{I}\cdot\mathbf{S} + g\beta\mathbf{B}_r(t)\cdot\mathbf{S} \tag{5.38}$$

This case can therefore be related to the discussion of Section 3.3, except that now the hyperfine interaction is isotropic as opposed to being axially symmetric as assumed in Equation (5.27). The analysis of the relaxation effects is then according to that presented in Section 3.3, where $\tau_C$ may be identified with the correlation time for spin fluctuations governed by the dipolar interaction and $\langle B_r^2 \rangle$ can be related to the mean square dipolar field.

Experimental data at 4.2 K in zero external magnetic field are shown in Figure 5.8. Analysis of the data shows that relaxation phenomena must be considered. However, the presence of a magnetically split pattern implies that the relaxation time must be long in comparison with the Larmor precession time $\tau_L$ (which corresponds to $A^{-1}$ in the present case). Here is therefore an example in which a resolved magnetic splitting is seen, even though the spin system is paramagnetic (*cf.* the third paragraph in Section 3.2).

There are two ways of interpreting the data. If it is assumed that $\tau_C \ll \tau_L$ the white noise approximation of case (i) in Section 3.3 applies. The spectrum in Figure 5.8 may then be fitted with Equation (5.32), where the slow relaxation regime $\tau_R > \tau_L$ applies. The results of fitting the data are indicated by the dashed lines. In the fits, the value of the hyperfine parameter $A$ has been constrained to be 699 MHz, as determined from prior electron nuclear double resonance measurements. It is clear that the white noise approximation leads to a discrepancy with the data. The reason for this is not hard to find. Analysis of the data shows that $\tau_R^{-1} \sim 212$ MHz. This, together with a direct lattice sum value of $\langle B_r^2 \rangle$, yields from Equation (5.31) a value for $\tau_C^{-1}$ of around 1500 MHz. This result shows that $\tau_C$ is comparable to $A^{-1}$ and hence the assumption of the noise being white must be discarded. We therefore have to consider a different approach to the interpretation (case (ii) of Section 3.3) which requires a more detailed theoretical analysis than covered here (see Dattagupta, Shenoy, Dunlap & Asch, 1977). Such an analysis leads to the solid lines in Figure 5.8. From this the spin–spin correlation time $\tau_C$ for the dipolar coupled spins can be estimated.

Fig. 5.8. The Mössbauer spectrum of $Cs_2NaYbCl_6$ obtained at 4.2 K. The dashed line represents the fit based on the white noise approximation. The solid line is the fit obtained from the improved theory of Dattagupta *et al.* (1977).

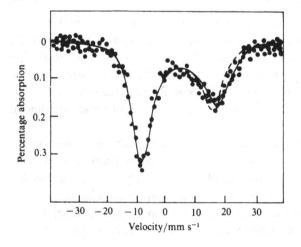

### 3.5  Comparisons with other techniques

The relaxation effects as probed via the Mössbauer hyperfine interaction can be investigated by other methods which can be regarded as complementary to Mössbauer spectroscopy. The techniques of perturbed angular correlation of gamma rays (PAC) and positive muon spin rotation ($\mu$SR) spectroscopy will be briefly discussed here. A more detailed review can be found in Dattagupta (1981).

The PAC experiment makes use of a nucleus which decays through successive gamma radiations $\gamma_1$ and $\gamma_2$. The observation of $\gamma_1$ in a fixed direction determines the spin orientation of the nucleus and the subsequent radiation $\gamma_2$ then has a definite angular correlation with respect to the direction of $\gamma_1$. If in the intermediate state of the gamma-ray cascade the nucleus is under the influence of a magnetic hyperfine interaction, the second radiation $\gamma_2$ carries with it information on this interaction. Hence the relaxation of the environment within the lifetime of the intermediate state of the nucleus manifests itself in a time-dependent magnetic hyperfine interaction.

Another method of orientating an emitter in its initial state and then observing the angular correlation of the subsequent radiation is used in $\mu$SR spectroscopy. First a pion decays into a muon and a mesonic neutrino. This process determines the spin polarisation of the muon at the time of its creation. Subsequently after a mean lifetime of approximately $2.2 \times 10^{-6}$ s the muon decays into a positron, a leptonic neutrino and a mesonic antineutrino.

Now the $\mu$SR experiment involves the determination of the spin polarisation of the muon at the time of its decay by observing the direction of emission of the positron. The situation is therefore analogous to the PAC case. Since the spin polarisation of the muon is known precisely at the time of its creation, any change in this produced as a result of the magnetic hyperfine interaction between the muon and its environment can be detected by measuring the direction of emission of the positron. The PAC and the $\mu$SR experiment are both performed as a function of time whereas the Mössbauer spectra are recorded as a function of the hyperfine interaction frequency (or energy).

It is important to note that in the context of relaxation studies a combination of Mössbauer spectroscopy and the PAC and $\mu$SR methods allows a given problem to be approached over a range of experimental timescales. For example, it may be recalled that the experimental timescales in Mössbauer spectroscopy are the lifetime of the excited state of the nucleus $\tau_N$ and the Larmor precession time $\tau_L$, while in the case of PAC the appropriate timescale is the lifetime of the intermediate nuclear state and for $\mu$SR it is the lifetime of the muon, approximately $2.2 \times 10^{-6}$ s. Therefore the same physical phenomenon, such as the diffusion of an interstitial atom, which would give rise to fluctuating magnetic hyperfine interactions, can be usefully investigated by a combination of different techniques.

**Acknowledgement**

This chapter was prepared while the author was a fellow of the Alexander von Humboldt Foundation at the Solid State Research Institute of the Kernforschungsanlage, Jülich, Federal Republic of Germany.

# 6

# The dynamics of nuclei studied by Mössbauer spectroscopy†

E. R. BAUMINGER &
I. NOWIK

## 1    Introduction

The purpose of this chapter is to review the application of Mössbauer spectroscopy to the study of dynamics. The main emphasis will be on the new areas in which significant advances have been made in recent years, such as the study of diffusive processes, the influence of motion on lineshape, time-dependent studies and dynamics at phase transitions. Rather less attention will be devoted to areas such as $f$-factors in solids, the Goldanskii–Karyagin effect and temperature shift, which have been studied more extensively in the past. The trends towards future areas for this research will also be considered, together with an evaluation and comparison between Mössbauer spectroscopy and other methods for investigating these phenomena.

No attempt will be made to develop the formal theory of dynamics in this chapter. General physical arguments will be employed to describe the phenomena and the main theoretical results to be used in interpreting the experimental data will be summarised. Representative experimental data and results will be presented in the various sections.

The phenomenon of the recoil-free resonant absorption of gamma rays is the result of the special dynamics of nuclei in solids. This fact was recognised by Mössbauer, who interpreted his apparently anomalous observations in terms of lattice dynamics of a similar nature to those associated with neutron scattering (Mössbauer, 1958). He found that the recoil momentum of the recoiling nucleus is shared by many nuclei of the crystal and thus the recoil energy is extremely small. This cooperative

† The untimely death of Professor S. G. Cohen made it impossible for him to write this review. We have written this chapter in the midst of our grief over his loss and in the hope that it will contribute to preserving his memory as an outstanding scientist and teacher, a warm human being and a close personal friend.

absorption of the recoil momentum arises because the whole crystal is a single quantum system in which the individual nuclei are coupled to each other and participate in cooperative vibrational modes. Thus, the first Mössbauer effect studies were actually investigations into lattice dynamics, which showed that the temperature dependence of the recoil-free fraction corresponds to the changes of the relative population of the various vibrational modes with temperature. In particular it was shown that, contrary to previous expectations, resonance absorption increases as the temperature decreases. This is due to the fact that at low temperatures only low-energy phonons are available and there is therefore only a low probability for the transfer of the nuclear recoil energy to the lattice. It turns out that if $f$ is the recoil-free fraction, then $-\ln f$ is proportional to $\langle x^2 \rangle$, the mean-square displacement of the absorbing nucleus in the gamma-ray direction. Details about the temperature dependence of the recoil-free fraction are given in Section 2.1.

Several years after the discovery of the Mössbauer effect it was suggested that the energy of the resonance line should depend on lattice dynamics and thus also on temperature (Pound & Rebka, 1959; Josephson, 1960). This so-called thermal shift can be looked upon as resulting from the second-order Doppler shift, or from the change in kinetic energy of the absorbing nucleus due to its change in mass. This shift in relative energy is equal to $-\frac{1}{2}\langle v^2 \rangle / c^2$ where $\langle v^2 \rangle$ is the mean-square velocity of the nucleus and $c$ is the speed of light. A study of the temperature dependence of the intensity and position of a Mössbauer absorption line will therefore yield the temperature dependence of both the mean-square vibrational amplitude and the mean-square velocity of the absorbing nucleus. A great advantage of Mössbauer spectroscopy over other techniques is that it also yields the dynamics of the Mössbauer nucleus when it is a probe such as a dilute impurity in a host crystal. Under such conditions the dynamics of the impurity are frequently very different from that of the host material and the experimental study can be used to check various models for local modes. This is considered further in Section 2.2.

Since the Mössbauer recoil-free fraction is related to the mean-square displacement in the direction of the gamma radiation, it can provide information on the anisotropy in vibrational modes. For nuclei in atoms occupying crystal sites of less than cubic symmetry, such as on surfaces, in membranes, intercalated between layers of graphite or in any other conditions of low symmetry, the mean-square displacement will be strongly anisotropic. This anisotropy can be easily detected by Mössbauer measurements in which the angle between the gamma-ray beam and the local coordinates of the absorbing system, e.g. single crystal, surface, membrane or layer, is changed. Such studies have been performed and very large anisotropies in vibrational modes have been detected (Section 2.3).

When the absorbing system is randomly oriented, such as in polycrystalline materials, and gives a single-line Mössbauer spectrum, there will be no way to detect anisotropy in the mean-square displacement. However, if the spectrum of the absorbing nuclei displays resolved hyperfine structure with either quadrupole or magnetic splitting or both, the relative intensities of the various absorption lines again yield information on the anisotropy in $\langle x^2 \rangle$. This phenomenon, which was first discussed by Karyagin and Goldanskii and is called the Goldanskii–Karyagin effect (Karyagin, 1963; Goldanskii, Makarov & Khrapov, 1963), will be considered further in Section 2.4.

Since the positions and intensities of Mössbauer absorption lines give information on lattice dynamics, any changes in the lattice dynamics at a given temperature will be reflected in the corresponding Mössbauer spectra. Thus any phase transition which involves changes in lattice dynamics can be detected through Mössbauer spectroscopy. In a crystallographic phase transition the interatomic distances and the symmetry of the crystal change, leading to a change in the vibrational modes of the crystal, and hence a discontinuous jump in the recoil-free fraction and thermal shift of the Mössbauer spectrum. Other transitions, like ferroelectric and superconducting transitions, are associated with a soft phonon mode which affects the recoil-free fraction $f$. Magnetic transitions in crystals with a large magnetostriction also display changes in both $f$ and line positions at the phase transition temperature; either an increase or decrease in $f$ may be observed at this temperature. Finally, a solid–liquid or solid–glass transition gives rise to a sharp change in the recoil-free fraction, and in many cases a change in the general shape of the spectrum will also occur (see Section 3).

The effects of diffusion on the lineshape are discussed in Sections 4 and 5. Free diffusion mainly affects the Mössbauer linewidth, with the increase in linewidth being proportional to the diffusion coefficient. Limited or bounded diffusion, which may occur under special conditions in a crystal or in a big molecular system such as a biological system, causes the Mössbauer lineshape to become very unusual. The shape often appears like two superimposed lines, one of natural linewidth, corresponding to elastic absorption, and the other with a much larger linewidth, corresponding to quasi-elastic absorption.

Information similar to that obtained by Mössbauer studies concerning the lattice dynamics and general dynamics of nuclei can be obtained by a wide variety of other techniques. Some are macroscopic bulk measurements such as heat capacity or sound velocity and absorption measurements, while others are microscopic methods such as neutron scattering, X-ray scattering, Rayleigh scattering and Rayleigh–Mössbauer scattering. We shall examine each of these methods in comparison with the Mössbauer technique in Section 6. Some general conclusions concerning the study of the dynamics of nuclei by Mössbauer spectroscopy and a discussion of future trends are given in Section 7.

## 2    Dynamics of nuclei in crystals

The Mössbauer recoil-free fraction, the anisotropy in the recoil-free fraction and the thermal shift, when studied as a function of temperature, may all yield information on lattice dynamics which is as useful as that obtained from heat capacity or X-ray scattering measurements. One obvious advantage of the Mössbauer technique is the fact that it measures the vibrational modes of the Mössbauer probe, thereby yielding information on local modes which is unobtainable by other techniques.

### 2.1    *The recoil-free fraction*

The subject of the recoil-free fraction was initially discussed in Mössbauer's original paper (Mössbauer, 1958). Since then, many books and review articles (e.g. Nussbaum, 1966) have been devoted to describing experimental measurements and theoretical calculations of the recoil-free fraction $f$. Hence this discussion will be limited to some of the major aspects of the information obtainable from $f$-factor measurements. The measurement itself is relatively easy, since for thin absorbers the Mössbauer spectral area is proportional to $f$, and thus the relative change in $f$ with temperature is readily obtainable. With thick absorbers one can still obtain accurate values, although the analysis requires transmission integrals (Shenoy, 1973).

The fraction of gamma rays emitted or absorbed without recoil by a nucleus bound in a crystal is given to a good approximation by the expression:

$$f = \exp\left(-k^2 \langle x^2 \rangle\right) = \exp\left(-2W\right) \qquad (6.1)$$

Where $k$ is the wavevector of the gamma ray and $\langle x^2 \rangle$ is the mean-square displacement of the Mössbauer nucleus in the direction of the gamma ray. The exponent $2W$ is the Debye–Waller factor, which is also relevant to X-ray and neutron scattering line intensities. Thus a measurement of the recoil-free fraction at a particular temperature $f(T)$ yields the value of $\langle x^2 \rangle$ at this temperature, which can be compared with theoretical calculations of

$\langle x^2 \rangle$ within various models. The value of $\langle x^2 \rangle$ depends directly on the local density of phonon frequencies, the spectrum of which depends on the forces acting between the atoms in the crystal. Thus measurements of $f(T)$ can give information on interatomic forces and the lattice frequency spectrum. If $G(\Omega)$ is the normalised phonon frequency distribution function, then in the harmonic approximation $\langle x^2 \rangle$ is given by

$$k^2 \langle x^2 \rangle = \frac{\hbar k^2}{2M} \int_0^{\Omega_{\max}} \Omega^{-1} \coth \frac{\hbar \Omega}{2k_B T} \, G(\Omega) \, d\Omega \qquad (6.2)$$

where $M$ is the mass of the Mössbauer nucleus and $\Omega_{\max}$ is the maximum phonon frequency.

The distribution function $G(\Omega)$ is generally derived using a specific model. In the Einstein oscillator model there is only one frequency $\Omega_E$ and thus $G(\Omega) = \delta_K(\Omega - \Omega_E)$, where $\delta_K$ is the Kronekker delta function. Hence, defining an Einstein temperature $\theta_E$, given by $\hbar \Omega_E / k_B$, this gives

$$k^2 \langle x^2 \rangle = \frac{E_R}{k_B \theta_E} \coth (\theta_E / 2T) \qquad (6.3)$$

where $E_R = \hbar^2 k^2 / 2M$ is the recoil energy.

In the Debye model approximation, which is valid for cubic monatomic crystals, the normalised density of vibrational modes is given by $G(\Omega) = 3\Omega^2 / \Omega_D^3$, where $\Omega_D$ is the maximum or Debye frequency. This leads to

$$k^2 \langle x^2 \rangle = (6E_R / k_B \theta_D) \left[ \frac{1}{4} + \left( \frac{T}{\theta_D} \right)^2 \int_0^{\theta_D/T} y(e^y - 1)^{-1} \, dy \right] \qquad (6.4)$$

where $\theta_D$ is the Debye temperature, given by $\hbar \Omega_D / k_B$.

Although a real crystal will seldom follow these formulae they can serve as a first approximation. In many cases the function $G(\Omega)$ can be measured directly by neutron scattering and thus one can compare experimental $f(T)$ values with those predicted by Equation (6.2). In the simple approximation $f(T)$ is given in the high temperature limit $(k_B T \geq \hbar \Omega_{\max})$ by $\exp(-AT)$, where $A$ is a constant. If the spectral area is proportional to $f$, and is therefore given by $Cf$, where $C$ is another constant, a plot of the logarithm of the spectral area at high temperatures against temperature will give both $A$ and $C$. Any deviation from linearity will indicate a temperature dependence of $\Omega_{\max}$, or the presence of anharmonicity in the nuclear vibrations.

In molecular crystals each atom participates in two kinds of motions, one due to intramolecular forces and the other due to the motion of the whole molecule, with relatively weak intermolecular forces. Since the two kinds of motions are only weakly correlated, the total recoil-free fraction can be expressed as the product of the two corresponding recoil-free fractions. It is claimed that a good approximation for the intramolecular motion is the

Einstein model. The analysis of the recoil-free fraction in molecular crystals and in particular in big molecules, such as in organic crystals of biological interest, is quite complicated and is discussed in more detail in Section 5.

As mentioned before, the first study of $f(T)$ was carried out by Mössbauer (1958), and since then many systems have been studied. Mössbauer's results were theoretically analysed by Visscher (1960). The effective absorption cross-section for $^{191}$Ir as a function of temperature is shown in Figure 6.1, while in Figure 6.2 the temperature dependence of the Debye–Waller factor of tin metal is shown. Serious corrections to the Debye model are needed to fit the experimental observations. One correction is due to the expansion of the crystal, which causes $\theta_D$, the Debye temperature, to be temperature dependent. The other correction is due to the anharmonicity of the nuclear motion (Boyle, Bunbury, Edwards & Hall, 1961).

### 2.2    *Local modes*

It has been mentioned above that one of the major advantages of the Mössbauer technique, compared with other methods which measure the parameters of lattice dynamics, is its ability to measure local modes of the Mössbauer probe. Dilute Mössbauer probes can be introduced into

Fig. 6.1. The recoil-free fraction $f$ of the absorption of the 129 keV gamma ray of $^{191}$Ir in iridium metal as a function of source temperature $T$. The curve is the theoretical fit. (Visscher, 1960.)

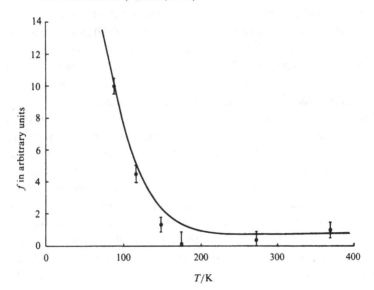

almost any lattice, either as a substitute or interstitially, and the local recoil-free fraction can then be measured. The values and temperature dependence of the recoil-free fraction will yield information on the local environment of the probe, on host–probe interactions, and on the lattice dynamics of the host.

The problem of the vibrational modes of an impurity nucleus in a host lattice concerned physicists long before the discovery of the Mössbauer effect (Maradudin, 1966). In Maradudin's review the problem of both isotopic impurities and other element impurities with masses either lighter or heavier than the host nuclei are discussed, together with the results expected from a Mössbauer measurement. The early experimental results shown in Figure 6.3 (Bryukhanov, Delyagin & Kagan, 1963) agree with the simple model relevant to an isotopic impurity. Under these circumstances the local vibrations of the probe are such that the recoil-free fraction can be expressed in terms of the mean-square deviation of the host lattice. At low temperatures

$$\langle x^2 \rangle_{\text{imp}} = \langle x^2 \rangle_{\text{host}} (m_{\text{host}}/m_{\text{imp}})^{\frac{1}{2}} \tag{6.5}$$

where $m_{\text{host}}$ and $m_{\text{imp}}$ are the atomic masses of the host and impurity respectively. If the lattice dynamics of the host can be approximated by a Debye model, then the recoil-free fraction of the impurity can also be

Fig. 6.2. Temperature dependence of the Debye–Waller factor $2W$ for metallic tin and its comparison with theory: (a) Debye model with $\theta_D = 142$ K, (b) allowing a temperature dependent $\theta_D$, and (c) adding anharmonicity contributions. (Boyle et al., 1961.)

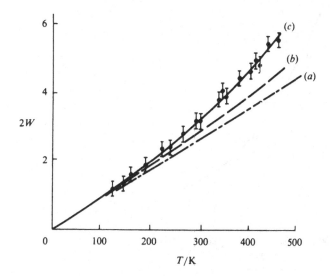

described by the same model, except that $\theta_D$ has to be replaced by $\theta'_D = \theta_D(m_{host}/m_{imp})^{\frac{1}{2}}$.

More recent experiments have shown that such a model is an extreme over-simplification. Impurities form local modes or resonance modes under a variety of conditions which depend on relative masses and relative interatomic interactions. A detailed theory has been worked out for the recoil-free fraction of an impurity (Mannheim, 1968). It is expressed in terms of the host lattice phonon frequency distribution function, the ratio of the masses of the host and impurity atoms and the ratio of the force constant between host atoms to that between host and impurity atoms. In Figure 6.4 the experimental results and the theoretical curves for $^{57}$Fe in vanadium are shown. Further details and recent experimental results for this model may be found in the Proceedings of the Indian National Science Academy: International Conference on the Applications of the Mössbauer Effect (1982), pp. 619–63.

Fig. 6.3. Temperature dependence of the recoil-free fraction of the absorption of the 24 keV gamma ray of $^{119}$Sn in tin nuclei bound in vanadium metal. (Bryukhanov *et al.*, 1963.)

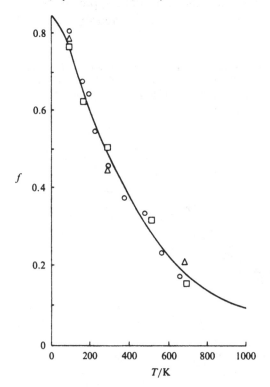

In several cases Mössbauer studies have revealed a recoil-free fraction with a very flat temperature dependence, which is significantly different from the predictions of either the Einstein or Debye models. Such results were mainly obtained with the Mössbauer probe atom bound within a volume large relative to the atomic size of the probe. They were analysed within a square well potential model, which yields a temperature-independent recoil-free fraction, since in a square well $\langle x^2 \rangle$ is bound for all states and is dictated by the width of the well, in contrast to the increasing width of the harmonic potential (Nussbaum, 1966).

### 2.3    Anisotropy in the recoil-free fraction

If the Mössbauer nucleus is located in a non-cubic site within the crystal it will have an anisotropic mean-square vibrational amplitude. The recoil-free fraction will then depend on the orientation of the gamma ray relative to the local coordinate axes of the absorbing nucleus.

Only cases in which the absorbing nucleus is located in a site with axial symmetry have been subjected to extensive experimental study and have been analysed in detail. In such cases the mean-square vibrational amplitude in the gamma-ray direction $\langle x_\gamma^2 \rangle$ is given by

$$\langle x_\gamma^2 \rangle = \langle x_\perp^2 \rangle + (\langle x_\parallel^2 \rangle - \langle x_\perp^2 \rangle)\cos^2 \theta \qquad (6.6)$$

where $\langle x_\perp^2 \rangle$ and $\langle x_\parallel^2 \rangle$ are the mean-square vibrational amplitudes

Fig. 6.4. Temperature dependence of the recoil-free fraction of the 14.4 keV gamma ray of $^{57}$Fe in iron nuclei bound in vanadium metal. The open circles represent the experimental points. The lines are theoretical predictions for various ratios of the force constant between host and impurity atoms, $\lambda'$, to that between host atoms, $\lambda$. (Mannheim, 1968.)

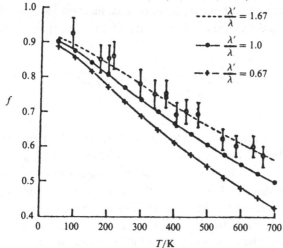

perpendicular and parallel to the axis of symmetry, and $\theta$ is the angle between the direction of the gamma ray and the axis of symmetry. The recoil-free fraction is then given by

$$f(\theta) = \exp\left[-k^2\langle x_\perp^2\rangle - k^2(\langle x_\parallel^2\rangle - \langle x_\perp^2\rangle)\cos^2\theta\right] \qquad (6.7)$$

where $k^2(\langle x_\parallel^2\rangle - \langle x_\perp^2\rangle)$ is a measure of the anisotropy.

Studies with single-crystal absorbers can give the details of $f(\theta)$. Mössbauer nuclei attached to a surface of a crystal, or embedded in other two-dimensional systems, e.g. Mössbauer nuclei intercalated into graphite, also exhibit large anisotropy in their recoil-free fraction. Several studies of single crystals and two-dimensional systems have been reported. In most cases the isotopes used for the studies were $^{119}$Sn and $^{57}$Fe (Herber & Maeda, 1980; Howard & Nussbaum, 1980).

### 2.4    The Goldanskii–Karyagin effect

The anisotropy in the recoil-free fraction can also be observed in a polycrystalline sample if the absorption spectrum displays resolved hyperfine structure (Karyagin, 1963). The anisotropy in the recoil-free fraction then leads to relative intensities of the various absorption lines which are different from those expected for a randomly oriented polycrystalline sample. If the major symmetry axis of a given crystallite is oriented at an angle $\theta$ to the direction of the gamma ray, then for an electric or magnetic dipole nuclear transition the intensity of the absorption lines will have an angular dependence given by $I_0(\theta) = f(\theta)(\frac{3}{2})\sin^2\theta$ for a $\Delta m_I = 0$ transition and $I_1(\theta) = f(\theta)(\frac{3}{4})(1 + \cos^2\theta)$ for a $\Delta m_I = \pm 1$ transition. For a randomly oriented sample the above expressions must be averaged over all possible values of $\theta$ and the value of the ratio of the absorption line intensities are then given by

$$\bar{I}_0/\bar{I}_1 = \left[\frac{1}{2}\int_0^\pi f(\theta)\frac{3}{2}\sin^3\theta\,d\theta\right]\bigg/\left[\frac{1}{2}\int_0^\pi f(\theta)\frac{3}{4}(1 + \cos^2\theta)\sin\theta\,d\theta\right]$$

$$(6.8)$$

This ratio is one if $f(\theta)$ is isotropic. If $f(\theta)$ has the form given by Equation (6.7), this ratio will depend only on the value of $k^2(\langle x_\parallel^2\rangle - \langle x_\perp^2\rangle)$ and a measurement of the relative intensities in an absorption spectrum gives the value of this parameter directly. If in addition the total absorption intensity is measured, the average recoil-free fraction can be estimated. If $a_i$ and $b_i$ are the nuclear transition probabilities for the $\Delta m_I = 0$ and $\Delta m_I = \pm 1$ transitions respectively, then the average recoil-free fraction, measured experimentally, will be given by

$$\bar{f} = \bar{I}_0\left(\sum_i a_i\right) + \bar{I}_1\left(\sum_i b_i\right)$$                        (6.9)

where the summations are over the $i$ absorption lines of each type, and where $\bar{I}_0$ and $\bar{I}_1$ are given by Equation (6.8). The combined measurements of $\bar{I}_0/\bar{I}_1$ and $\bar{f}$ give $\langle x_\parallel^2\rangle$ and $\langle x_\perp^2\rangle$ directly. The above procedure is valid only if the sample contains a genuinely random orientation of crystallites, with no texture effects (Pfannes & Gonser, 1973) and the absorbers are thin enough to avoid distortions of the shape and intensity ratios of the various absorption lines (Shenoy & Friedt, 1973).

The effect of an anisotropic recoil-free fraction on the relative intensities of absorption lines in hyperfine split spectra has been observed and interpreted in $^{57}$Fe (Goldanskii, Makarov & Khrapov, 1963) and the effect has subsequently been observed in almost all Mössbauer isotopes. Particularly large effects have been seen with $^{151}$Eu in $Eu_2Ti_2O_7$ (Armon *et al.*, 1973; Bauminger *et al.*, 1974) and examples are shown in Figure 6.5.

Fig. 6.5. Mössbauer spectra of $^{151}$Eu in $Eu_2Ti_2O_7$. The dashed lines are theoretical curves with an isotropic recoil-free fraction. The full lines are computer fits to the experimental spectra assuming an anisotropic recoil-free fraction. (Bauminger *et al.*, 1974.)

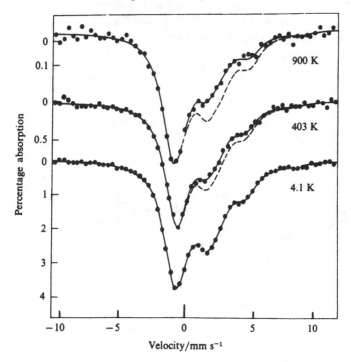

Recent studies of $^{151}$Eu in EuRu$_2$Ge$_2$ (Nowik & Felner, 1984) where the europium ion is divalent and is exposed to the largest electric field gradient ever observed in any rare earth metallic system, yield the temperature dependence of $f_{\parallel}$ and $f_{\perp}$ shown in Figure 6.6.

## 2.5    Thermal shift

The thermal shift of Mössbauer lines arises from the second-order Doppler effect (Pound & Rebka, 1959). If $\langle v^2 \rangle$ is the mean-square velocity of the Mössbauer nucleus in the crystal, then the relative shift in energy of the Mössbauer line will be given by $\Delta E/E_0 = -\langle v^2 \rangle/2c^2$. In the Debye model approximation this leads to

$$\frac{\Delta E}{E_0} = -\frac{9k_B\theta_D}{4Mc^2}\left[\frac{1}{4} + 2\left(\frac{T}{\theta_D}\right)^4 \int_0^{\theta_D/T} \frac{y^3}{e^y - 1}\,dy\right] \qquad (6.10)$$

In the more general case $\langle v^2 \rangle$ can be expressed in terms of the phonon frequency distribution function $G(\Omega)$ which can be measured experimentally

$$\langle v^2 \rangle = \frac{3\hbar}{2M} \int_0^{\Omega_{max}} \Omega \coth \frac{\hbar\Omega}{2k_BT}\, G(\Omega)\, d\Omega \qquad (6.11)$$

In the case of a Mössbauer impurity in a host crystal, $G(\Omega)$ of the host should be replaced by $G_{imp}(\Omega)$ of the impurity which is related to $G(\Omega)$ by the ratio of masses of the host and impurity atoms, and the ratio of impurity–host to host–host interaction parameters (Mannheim & Simopoulos, 1968). This is the same procedure as outlined for the calculation of the mean-square displacement $\langle x^2 \rangle$ in Section 2.2.

Fig. 6.6. Temperature dependence of the anisotropic recoil-free fraction of the absorption of the 21 keV gamma ray of $^{151}$Eu in EuRu$_2$Ge$_2$. (Nowik & Felner, 1985.) The solid lines are Debye model theoretical curves.

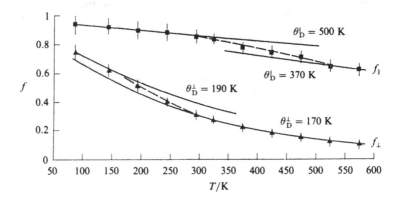

The thermal shift may be obscured by additional shifts associated with the isomer shift. The sensitivity of the thermal shift to the particular lattice dynamics parameters of the crystal is much less than that of the recoil-free fraction. In particular, at high temperatures the thermal shift $\Delta E(T)/E_0$ is model independent and linear with temperature, and is given by $-3k_B T/2Mc^2$. Nevertheless many experimental studies have been reported from which the various lattice dynamics parameters can be determined and compared with those obtained from the recoil-free fraction. A combined measurement of $\Delta E(T)/E_0$ and $f(T)$ can also yield the zero point mean-square velocity, which is of significance to studies of superconductivity (Kitchens, Craig & Taylor, 1970; Kimball, Taneja, Weber & Fradin, 1974).

## 2.6 Comparison with heat capacity measurements

The information obtained from Mössbauer spectroscopic studies of the temperature dependence of the recoil-free fraction is quite similar to that obtained by bulk measurements of the heat capacity. In the case of monatomic crystals, if $G(\Omega)$ is the normalised phonon spectrum, then the heat capacity at constant volume $C_v$ is given by

$$C_v(T) = 3Nk_B \int_0^{\Omega_{max}} \frac{(\hbar\Omega/k_B T)^2 \exp(\hbar\Omega/k_B T)}{[\exp(\hbar\Omega/k_B T) - 1]^2} G(\Omega) \, d\Omega \qquad (6.12)$$

where $N$ is the number of atoms. A clear comparison between $C_v(T)$ and $-\ln f(T)$ can be obtained from the intermediate- to high-temperature ($\hbar\Omega/k_B T < 6$) expansions (Nussbaum, 1966). Such a comparison shows that for $-\ln f(T)$ the dominant term is proportional to $\langle \Omega^{-2} \rangle$, while the dominant term for heat capacity is proportional to $\langle \Omega^2 \rangle$, where the frequency moments $\langle \Omega^n \rangle$ are given by $\int_0^{\Omega_{max}} \Omega^n G(\Omega) \, d\Omega$. Thus $f(T)$ and $C_v(T)$ have different sensitivities to low- or high-frequency regions of the phonon spectrum. Measurements of these quantities are therefore essentially complementary in nature.

The techniques of neutron diffraction, X-ray scattering and Rayleigh–Mössbauer scattering also yield information on the dynamics of nuclei. A detailed comparison between these techniques and Mössbauer spectroscopy has been given recently by Mössbauer (1983). X-ray diffraction, Rayleigh–Mössbauer scattering and elastic neutron scattering are all processes in which the scattering nucleus must stay recoil free. These processes therefore have cross-sections which are proportional to the recoil-free fraction and thus yield the same information as that obtained by Mössbauer measurements. Neutron scattering has the added advantage that inelastic scattering can also be measured and the lattice phonon spectrum $G(\Omega)$ can be determined directly. More details about the comparison between other scattering techniques and Mössbauer spectroscopy are given in Section 6.

## 3    Dynamics associated with phase transitions

The possible contributions of the Mössbauer technique to the study of phase transitions has been outlined previously (Shenoy, 1973). Almost all phase transitions cause changes in the lattice dynamics of the crystal and these changes can be studied through measurements of the recoil-free fraction, its anisotropy and the second-order Doppler shift. The phase transition itself is in many cases also observed through changes in the hyperfine interaction parameters.

### 3.1    Structural transitions

Crystallographic, order–disorder and ferroelectric phase transitions are phenomena which are all associated with the motion of the atoms in the crystal to a new equilibrium state. The phonon spectra above and below the transition temperature are consequently different and hence all the relevant Mössbauer parameters change. At these phase transitions discontinuities in the temperature dependence of the recoil-free fraction and of the thermal shift are therefore expected. The effect on the thermal shift may be masked by changes also occurring in the isomer shift. If the transition is associated with the softening of some lattice modes, as for example in ferroelectrics, a minimum in the recoil-free fraction at the phase transition should be observed (Shenoy, 1973). The phase transition itself is usually also made evident by sharp changes in the quadrupole splitting. A superconducting phase transition may also be associated with changes in the phonon spectrum and should therefore also be observable by Mössbauer measurements.

### 3.2    Magnetic ordering and spin reorientation transitions

Magnetic ordering phenomena are easily observable by Mössbauer spectroscopy and the technique is one of the major microscopic methods for the study of magnetic phenomena (see Chapter 4). However, little attention has been given to whether the Mössbauer technique can reveal changes in lattice dynamics associated with magnetic ordering. In particular, when a crystal exhibits strong magnetostriction behaviour, the magnetic ordering will introduce changes in the lattice dynamics, which have been observed by sound velocity measurements. In those crystals in which magnetostriction phenomena are observable, unusual changes in the recoil-free fraction and its anisotropy, and in the thermal shift, may be expected at the magnetic phase transition temperature. In systems in which the easy axis of magnetisation becomes reorientated at a certain temperature, changes in the Mössbauer parameters associated with lattice dynamics may also be observed.

The idea that Mössbauer spectroscopy should be sensitive to changes in lattice dynamics parameters at the magnetic phase transition was introduced by Bashkirov & Selyntin (1968), who showed that, within certain approximations, the recoil-free fraction and the thermal shift at the transition temperature can be expressed in terms of the magnon–phonon coupling strength. Another approach was introduced by Feder & Nowik (1979), who calculated the recoil-free fraction and thermal shift for a magnetic Debye solid with strong magnon–phonon coupling. Measurements have been performed on $RFe_2$ and $RCo_2$ systems (R = rare earth) which display these phenomena, and some of the results are shown in Figure 6.7.

The results show that at the magnetic phase transition a decrease or increase in the recoil-free fraction can be observed. The magnon–phonon coupling causes the phonon frequency distribution to shift to lower frequencies and to broaden. These two phenomena have opposite effects on the recoil-free fraction. While the frequency shift obviously produces a phonon softening and a consequent decrease in $f$, the broadening of the frequency distribution is effectively like a phonon hardening and leads to an increase in $f$. It can happen that in the same crystal the mean-square displacement in one direction increases while in another direction it decreases (Feder & Nowik, 1979).

The effect of magnetic ordering on the thermal shift is difficult to detect, since the ordinary isomer shift is also affected, in most cases to a greater extent than the thermal shift. Thus in most systems any line shift which has been observed at the ordering temperature could be explained in terms of electronic rearrangement effects which contribute to the isomer shift.

Sound velocity measurements on crystals with large magnetostriction provide evidence for changes in the lattice dynamics at the spin reorientation transition. Mössbauer studies of such systems show that there are changes in the recoil-free fraction at the spin reorientation transition (Seh & Nowik, 1981).

### 3.3 *Glass–liquid and solid–liquid transitions*

Glass-forming substances are characterised by a glass transition temperature $T_g$, above which the material is in a supercooled liquid state and below which a glass is formed. This transition is associated with a pronounced change in the heat capacity and should also give rise to changes in Mössbauer absorption parameters such as the recoil-free fraction, the linewidth and the quadrupole splitting.

An early review of glass transitions and their observation by Mössbauer spectroscopy has been given by Ruby (1973). The only difference between a

Fig. 6.7. Mössbauer parameters for $^{57}$Fe in DyFe$_2$ as a function of temperature, showing the decrease in the recoil-free fraction at the magnetic phase transition temperature $T_{ord}$. (Feder & Nowik, 1979.)

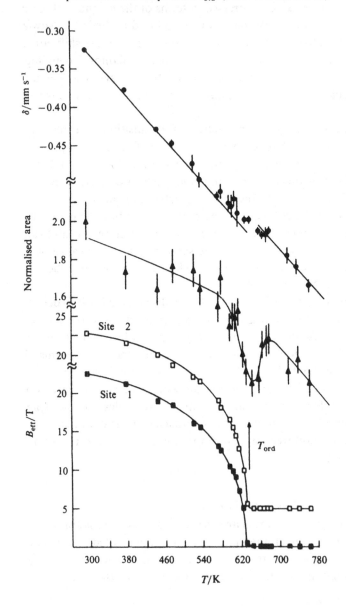

glass and a liquid is in the relevant relaxation times, which in the glass state are very long, with very little or no atomic rearrangements. Close to $T_g$ and above, the relaxation times become short and the atoms begin to diffuse and rearrange within times appropriate for Mössbauer spectroscopic observations. The mean-square deviation of each atom then increases sharply and the recoil-free fraction therefore decreases. The diffusive motion of the atoms also contributes to broadening of the Mössbauer line, from which the jumping rate or diffusion constant can be derived, as discussed in Section 4.

The recoil-free fraction as a function of the temperature for glass-forming $FeCl_2$ in methanol is shown in Figure 6.8. A sharp decrease in $f$ at $T_g$ is observed; while the increase in $f$ at higher temperatures is due to crystallisation (Simopoulos, Wickman, Kostikas & Petrides, 1970). The solid to liquid transition has also been studied in detail (Boyle *et al.*, 1961) and the decrease in $f$ and the increase in absorption linewidth near the melting point have been very clearly observed (Figure 6.9).

**4     Diffusion in liquids and solids**

*4.1     Introduction*

The transport of matter within a solid or liquid is described by the diffusion constant $D$, which is temperature dependent. In liquids, the translational diffusion constant is closely related to the viscosity coefficient

Fig. 6.8. Temperature dependence of the recoil-free fraction of the absorption of the 14.4 keV gamma ray of $^{57}Fe$ in glass-forming $^{57}FeCl_2$ in methanol. (After Simopoulos *et al.*, 1970.)

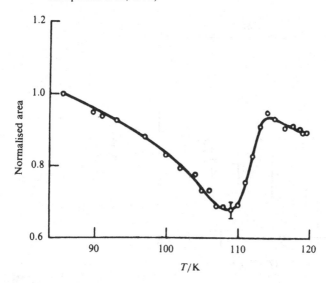

of the fluid $\eta$, through the expression $D = k_B T/6\pi r\eta$, where $r$ is the radius of the diffusing particle. Diffusion may be described as taking place by means of elementary diffusion jumps. In this model the atom is considered to vibrate around its site for a mean time interval $\tau_v$ and then translates to another site during a mean time interval $\tau_t$. In liquids the diffusion is a quasi-continuous process, with $\tau_v \ll \tau_t$, while in solids the diffusion is regarded as taking place by sudden jumps from one site to another, with $\tau_v \gg \tau_t$. The diffusion might be free or might be confined to a certain region, in which case it is described as bounded diffusion or diffusion within a cage. Diffusion constants in solids are usually measured by the tracer sectioning technique, which involves a radioactive isotope being electroplated on the

Fig. 6.9. (*a*) The temperature dependence of the recoil-free fraction of the absorption of the 24 keV gamma ray of $^{119}$Sn in tin metal near the melting point $T_m$. (*b*) The temperature dependence of the normalised diffusion linewidth $\Gamma_D/\Gamma$ of the 24 keV gamma ray of Sn$^{119}$ in tin metal near the melting point. The corresponding diffusion coefficient is given on the right-hand axis. (Boyle *et al.*, 1961.)

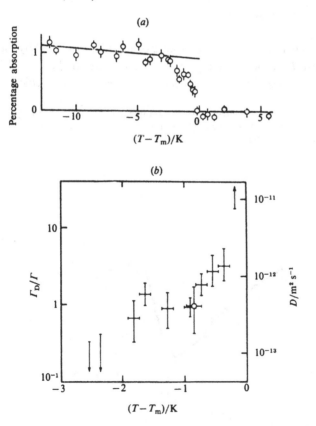

metal whose diffusion constant is to be determined. The metal is held at a certain temperature while the radioisotope diffuses into the material. Subsequently the sample is sliced and the diffusion profile is determined. This is a macroscopic measurement of the diffusion constant and does not provide much information on the microscopic process. Furthermore it is a destructive procedure and the diffusion constant at each temperature has to be determined on a different sample.

Several review articles on diffusion measurements using the Mössbauer technique have been published recently (Flinn, 1980; Mullen, 1982; Petry, 1983) and most of the relevant references for the original works can be found therein. In contrast to the tracer section method, measurements using Mössbauer spectroscopy do not destroy the samples and thus the temperature dependence of the diffusion can be measured on a single sample with the same impurity content. The Mössbauer measurement is also sensitive to changes in the immediate environment of the probe atom and may therefore give deeper insight into the microscopic mechanism of diffusion. The Mössbauer spectrum depends on the direction of the diffusive jumps relative to the gamma-ray beam and therefore contains information on the geometry of diffusion if the measurements are made on single crystals. In polycrystalline samples the anisotropy of the diffusion gives rise to different widths for the absorption lines corresponding to different hyperfine components (Goldanskii & Korytko, 1973).

Diffusion manifests itself in the profile of the Mössbauer spectrum through the broadening and changes in shape of the absorption lines and also through relaxation effects due to the variation of the hyperfine interactions with time. As diffusion in solids can be most readily detected in the Mössbauer spectra at elevated temperatures close to the melting point, it will only be observed for those Mössbauer nuclei which have a relatively large recoil-free fraction at these temperatures. Almost all of the work on diffusion using Mössbauer spectroscopy has therefore been carried out with $^{57}$Fe and $^{119}$Sn, with only a very few studies using $^{181}$Ta.

The range of timescales over which diffusion can be observed in Mössbauer spectroscopy is determined by the lifetime of the excited state of the Mössbauer nucleus and the Larmor precession time of the hyperfine interactions. For $^{57}$Fe the times involved are in the range between $5 \times 10^{-7}$ and $5 \times 10^{-9}$ s and the diffusional line broadening is equal to the natural linewidth when the diffusion constant $D$ is of the order of $10^{-15}$ m$^2$/s in a liquid.

The concept of diffusional broadening was originally developed for the incoherent quasi-elastic scattering of slow neutrons from moving particles. The resolution in this situation is determined by the instrumentation and allows measurements of diffusion constants of $10^{-12}$ m$^2$/s and greater. This technique therefore gives effects at higher temperatures than Mössbauer spectroscopy. Since the diffusion line broadening in solids is proportional to $k^2D$, where $k$ corresponds to the momentum transfer and, because in Mössbauer spectroscopy the wavevector $k$ is fixed, the diffusion broadening can only be measured in a small temperature range, due to the sharp decrease of the recoil-free fraction. In neutron scattering the momentum transfer corresponds to between zero and about 20 nm$^{-1}$, and the diffusional broadening can therefore be measured over a wide range of temperatures. A comparison of Mössbauer spectroscopy and neutron scattering measurements for diffusion of Fe and Co in Zr has been described by Petry (1983).

The analysis of Mössbauer spectra affected by diffusion is based on an early work by Singwi & Sjölander (1960). The absorption cross-section $\sigma_a$ is given in this theory by the space and time Fourier transform of the nuclear decay factor $\exp(-t/\tau_N)$ times the self-correlation function $G(r, t)$, which is the probability of finding the Mössbauer nucleus which was at the origin at time $t=0$, at a position $r$ at time $t$.

$$\sigma_a(k, \omega) = \frac{\sigma_{max}}{4\tau_N} \int\!\!\int \exp\left\{i[\mathbf{k} \cdot \mathbf{r} - (\omega - \omega_0)t] - t/2\tau_N\right\} G(r, t)\, dr\, dt$$

$$(6.13)$$

where $\tau_N$ and $\hbar\omega_0$ are the mean lifetime and energy of the nuclear excited state, $k$ is the wavevector of the gamma ray ($k = 2\pi/\lambda$), and $\sigma_{max}$ is the maximum cross-section. When no diffusion is present, $G(r, t)$ represents the vibrational motion of the nucleus and the time integration gives a Lorentzian lineshape with the natural width and the space integration gives the $f$-factor. When diffusion occurs $G(r, t)$ has to be specified. The Singwi and Sjölander theory assumes the diffusive jumps to be isotropic and long relative to $\lambda$, with no correlations between successive jumps and complete independence of the vibrational and translational components of the motion. In solids the theory has to be modified to take into account correlation between successive jumps. Although some doubt as to the validity of the assumption of the independence of the vibrational and translational motions has been expressed (Mullen, 1982), this assumption has been retained in all later modifications of the theory.

## 4.2    *Diffusion in liquids*
### 4.2.1    *Suspended particles in viscous liquids and solutions*

In liquids the diffusion is isotropic and is a quasi-continuous process. A consideration of the Singwi and Sjölander theory and the quasi-continuous motion leads to a calculated line with almost Lorentzian shape and with a width given by $\Gamma + 2\hbar k^2 D$, where $\Gamma$ is the natural linewidth and $2\hbar k^2 D$ is the diffusion broadening $\Gamma_D$. The calculated deviation from a Lorentzian lineshape is small and very difficult to observe (Flinn, 1980).

Suspended particles in a liquid will be subject to rotational as well as translational diffusion. The translational diffusion coefficient $D_t$ of a particle of radius $r$ undergoing small jumps with a mean-square jump distance $\langle d^2 \rangle$, is given by $(\frac{1}{6})\langle d^2 \rangle / \tau_v$, while the rotational diffusion, coefficient $D_r$ is given by the Stokes approximation as $D_t/r^2$. Since $r^2$ is usually much greater than $\langle d^2 \rangle$, the rotational diffusion affects the Mössbauer spectrum much less than the translational diffusion. A quantitative estimate of the effect of rotation on the width of the absorption line is given in Section 5. Rotational diffusion may also smear out the quadrupole splitting in a Mössbauer spectrum by relaxation effects related to the change in orientation of the electric field gradient, but this effect will only be observed in cases where $r^2$ is of the order of $\langle d^2 \rangle$ (Dattagupta, 1976).

The first Mössbauer diffusion experiments were performed on [57]Fe ions dispersed in liquid glycerol (Bunbury, Elliot, Hall & Williams, 1963; Craig & Suttin, 1963) and were subsequently extended to the examination of the diffusion of colloidal particles in viscous organic liquids and to Sn[119] probes in glycerol (Mullen, 1982). In these measurements Lorentzian lineshapes were observed and the linear relationship between the line broadening and the diffusion constant was confirmed. Using the relation $\Gamma_D = 2\hbar k^2 D = \hbar k^2 k_B T / 3\pi r \eta$, the radius $r$ of the colloidal particles in the solutions was determined, and found to agree with those measured by scanning electron microscopy. It therefore seems that the original theory of Singwi and Sjölander is satisfactory for most cases of colloidal particles investigated. Diffusion broadened Mössbauer lines of non-Lorentzian shape have been reported for $SnO_2$ particles in silicon grease and for $Fe(C_5H_7O_2)_3$ dissolved in 1,3-propanediol, implying that in these cases either rotational diffusion is the main mechanism of diffusion or the self-correlation function has to be modified to give non-Lorentzian lineshapes (Heilmann, Olsen & Jensen, 1974).

### 4.2.2    Liquid crystals

Ordered solute monocrystals can be obtained by dissolving iron- or tin-bearing molecules into liquid crystalline materials and cooling the materials in a magnetic field. In liquid crystals the molecules order in layers in which they orient themselves in preferred directions. By measuring the temperature and angular dependence of the $f$-factor and the linewidth, differences between the liquid crystal phase and the solid state can be detected and information about anisotropic diffusion can be obtained. In accordance with other measurements it was found that translation normal to the layers is more hindered than translation within a layer. Below the glass transition temperature, $f(T)$ shows the normal solid-like behaviour, with a dependence on orientation. Above the glass transition temperature, $f(T)$ deviates significantly from the Debye approximation and a rapid drop in the recoil-free fraction, which is characteristic for liquids, is observed. Angle dependent line broadening is only observed over a quite limited temperature range (La Price & Uhrich, 1979).

### 4.3     *Diffusion in solids*

In order to apply the Singwi and Sjölander theory to solids it must be modified to take into account the diffusion possibilities determined by the crystal structure, since the directions in which an atom can jump depend on the local environment and the diffusive jumps may therefore not be isotropic. Generally a diffusing atom can only move into a vacant neighbouring site, and there is therefore a substantial probability that the jumping atom will subsequently return to its original site. The appropriate correction, known from the theory of bulk diffusion as the Bardeen–Herring correlation factor, has been used in a number of papers to calculate the Mössbauer lineshape (Wolf, 1977). The diffusion line broadening in solids is given by $\Gamma_D = 2\hbar[1 - \alpha(k)]/\tau_v$, where $\alpha(k)$ is an anisotropy factor which has been calculated specifically for simple crystal structures, and is actually the Fourier transform of the probability that the Mössbauer atom will jump to a certain point (Chudley & Elliot, 1961). The diffusion constant $D$ is proportional to $1/\tau_v$ and so the broadening is always proportional to the diffusion constant. For cubic metals for instance, $D = a_0^2 f_c/6\tau_v$, where $a_0$ is the nearest-neighbour distance, and the Bardeen–Herring correlation factor $f_c$ reduces the line broadening predicted by the Singwi and Sjölander theory by about a factor of 2. In a number of measurements of iron in dilute solutions in cubic metals the line broadening is consistent with bulk diffusion data if these corrections are taken into account (Flinn, 1980). Due to the anisotropy in $\alpha(k)$, the predicted lineshape for polycrystalline materials is cusp-like (Dibar-Ure & Flinn, 1973). This deviation from Lorentzian lineshape is small and usually difficult to observe. Using strong radioactive doped single-crystal sources. The emission cross-section high-temperature furnace system a cusp-like spectral profile has been observed for $^{57}Co$ in iron stabilised into the bcc phase. This is shown in Figure 6.10 (Mullen, 1980).

By measurements on single crystals it is possible to learn more about the diffusion jump vector $r$. Mantl, Petry, Schroeder & Vogl (1983) demonstrated this for the diffusion of iron in aluminium single crystals. The measurements on single crystals were made possible by the preparation of strong radioactive doped single-crystal sources. The emission cross-section from a diffusing Mössbauer nucleus in a crystal in which the Mössbauer atom can jump to $n$ nearest sites, can be written as

$$\sigma_e(k, \omega) = f_v \frac{1}{2\pi\hbar} \int_{-\infty}^{+\infty} \exp\left[-i(\omega - \omega_0)t\right.$$

$$\left. - (\Gamma t/2\hbar)\right] \sum_i \exp(i\mathbf{k} \cdot \mathbf{r}_i) G(r_i, t)\, dt \quad (6.14)$$

where the sum covers all jump vectors $r_i$ to the $n$ possible sites of the diffusing atom, $f_v$ is the $f$-factor which takes account of the fast vibrational motion around the temporary equilibrium position, and $G(r_i, t)$ is the self-correlation function. The diffusion broadening is given in this case by

$$\Gamma_D = 2\hbar n v\left[1 - \sum_i P(r_i) \exp(i\mathbf{k} \cdot \mathbf{r}_i)\right] \quad (6.15)$$

where $P(r_i)$ is the probability that the Mössbauer atom jumps to site $r_i$, and $v$ is the jump frequency. The linewidth depends on the angles $\theta_i$ between $k$ and $r_i$. This anisotropic line broadening was observed by Mantl *et al.* (1983) in $^{57}$Fe in aluminium and is shown in Figure 6.11. By comparing this line broadening with calculations of $\Gamma_D$, it was concluded that iron diffuses into

Fig. 6.10. The Mössbauer spectrum of $^{57}$Co in polycrystalline bcc iron against a stainless steel absorber at 1283 °C. (Mullen, 1980.)

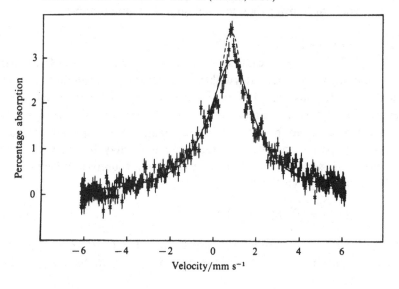

242    *The dynamics of nuclei*

Fig. 6.11. (a) Mössbauer spectra of $^{57}$Fe in an aluminium single crystal at 295 and 923 K for different orientations of the gamma-ray beam along the crystallographic directions. (b) Diffusional broadening $\Gamma_D$ at 923 K as a function of the observation direction specified by $\theta$, the polar angle; the azimuthal angle is 43°. The continuous line is the calculated anisotropy for self-diffusion and the broken line is for impurity diffusion. (Mullen, 1980.)

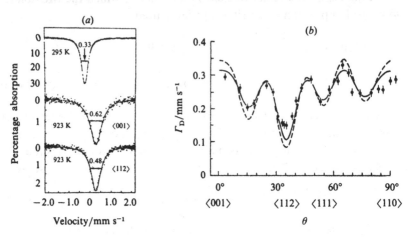

aluminium as a substitutional impurity by means of nearest-neighbour vacancies. The decrease in intensity at high temperatures, which is larger than the normal behaviour predicted by the Debye theory, was associated with Fe to vacancy binding, and the associated binding energy was estimated. From the temperature dependence of $\Gamma_D$ in different directions the diffusion coefficient of iron in aluminium was obtained and was found to be a factor of 5 lower than for aluminium in aluminium.

**4.4    Diffusion in a cage**

Diffusion in a cage is frequently described in terms of a Mössbauer atom jumping to $n$ discrete sites within a restricted region and thus jumping back to its initial position with a probability of $1/n$. As there is no phase shift between the parts of the gamma ray emitted from the same site, these should not give rise to any line broadening. Hence for any jump frequency a narrow line superimposed on different broadened lines may be expected. The emission cross-section for a Mössbauer atom which jumps to different sites within a cage is given by

$$\sigma_e(k,\omega)=f_v\sum_i I_i(k)L_i(\omega,\Gamma_i) \tag{6.16}$$

where $L_i$ represents a Lorentzian line of width $\Gamma_i=\Gamma+(\Gamma_D)_i$ and relative intensity $I_i(k)$. The intensity of the narrow line $I_0(k)$ has an angular dependence and is given by $I_0(k) = [(1/n) \sum_i \exp(i\mathbf{k} \cdot \mathbf{r}_i)]^2$. In cases where $n =$ 2, 4 or 6, relatively simple formulae for the shape of the Mössbauer spectrum have been obtained (Bläsius, Preston & Gonser, 1979). In the general case the broadened lines can be approximated by one broad quasi-elastic Lorentzian line of width $\Gamma + b\hbar/\tau_v$ where the parameter $b$ depends on the jump geometry, and $\tau_v$ follows the Arrhenius law $\tau_v = \tau_0 \exp(E_m/k_B T)$, where $\tau_0$ is a characteristic time and $E_m$ is the migration energy barrier.

Mössbauer spectra which display the phenomenon of diffusion in a cage have been observed in a number of investigations. The effect has been studied in aluminium single crystals doped with $^{57}Co$ and irradiated with 3 MeV electrons (Petry, Vogl & Mansel, 1980). Here the $f$-factor dropped sharply around 17 K and became anisotropic above this temperature. From comparison of the experimental anisotropy with results of model calculations of the emission probability, information on the arrangement of the jump sites was obtained. It was concluded that the Fe atom jumps in a cube-like cage. The dimensions of this cage and the thermally activated hopping frequency were determined from the temperature dependence of the linewidth of the broad line (Petry, Vogl & Mansel, 1982).

Line broadening observed in $^{57}Fe$ doped $TiH_x$ (Bläsius *et al.*, 1979) and the anomalous temperature dependence of the spectral areas in $^{57}Fe$ doped $VH_x$, $PdH_x$ and $NbH_x$ compounds (Probst, Wagner & Karger, 1980) have also been attributed to diffusion in a cage. It is believed that in these compounds the Mössbauer atom moves out of its equilibrium position and back as the diffusing hydrogen atom passes near by. The hydrogen appears to diffuse by thermally activated jumps over potential barriers between neighbouring interstitials. In this situation the Mössbauer spectrum of the $^{57}Fe$ doped hydrides should consist of a perfectly elastic line with the natural linewidth $\Gamma$ and a quasi-elastic broad line with a width $\Gamma + b\hbar/ [\tau_0 \exp(E_m/k_B T)]$. Eventually the quasi-elastic line becomes so broad that it is lost from the spectrum and this causes the anomalous decrease of the effective $f$-factor, which is assumed to be proportional to the spectral intensity. The quasi-elastic line has actually been only observed in $NbH_{0.78}$ near room temperature (Wordel & Wagner, 1984). In the other cases its existence and width have been inferred from the curves of recoil-free fraction against temperature. From the temperature dependence of the width of the quasi-elastic line, $E_m$ and $\tau_0/b$ can be deduced. From the temperature dependence of the intensity of the narrow line the displacement of the iron atom can be obtained and from the concentration dependence of this intensity the repulsive Fe—H interaction can be estimated. The activation energies obtained are in agreement with those obtained by neutron scattering and nuclear magnetic resonance (NMR) in $PdH_x$ and $NbH_x$.

Relaxation effects are only briefly mentioned here since they are dealt with in greater detail in Chapter 5. As diffusion in solids is usually observed only at elevated temperatures, its influence on hyperfine interactions is normally limited to quadrupole interactions and isomer shifts only. Whenever the Mössbauer atom is quasi-stationary on a cubic site the quadrupole interaction is caused by other diffusing atoms, as for example hydrogen or other solutes in a metal. At low temperatures a superposition of several quadrupole split spectra with different isomer shifts, corresponding to the different solute neighbours of the Mössbauer atom, may be observed. At high temperatures, due to fast relaxation, a single line with an average isomer shift may be seen and, at intermediate temperatures, broadened lines and intermediate shifts appear (Dattagupta, 1976). These effects have been seen in several experiments, as, for example, in $TaD_x$ and $TaH_x$, using $^{57}Fe$ probes (Heidemann, Wipf & Wortmann, 1978). Motional narrowing has also been observed using $^{181}Ta$ as the probe of hydrogen diffusion in tantalum metal (Wipf & Heidemann, 1980). When the Mössbauer atom is diffusing, the line is broadened due to diffusion and motionally narrowed by relaxation effects associated with the quadrupole splittings. In such cases the motional narrowing is much more difficult to observe.

## 5    The dynamics of diffusion in biological and macromolecular systems

### 5.1    *Introduction*

It is well known that biomolecules are not rigid systems and that their flexibility is often essential to their biological function. In various iron-containing proteins iron is at the active centre of the molecule and the dynamics involving the active centre may be of great importance. In the early Mössbauer studies of various proteins in solutions and in protein crystals, a sharp decrease of the recoil-free fraction around 210 K was observed (Parak & Formanek, 1971; Keller & Debrunner, 1980; Parak, Frolov, Mössbauer & Goldanskii, 1981). The temperature dependence of the recoil-free fraction indicated the appearance of new motional degrees of freedom above this temperature. More recent Mössbauer studies of iron-containing protein crystals (Cohen, Bauminger, Nowik & Ofer, 1981; Mayo, Parak & Mössbauer, 1981; Parak, Knapp & Kucheida, 1982; Bauminger *et al.*, 1983) reveal striking new phenomena relating to the dynamics of the matrix surrounding the iron in the proteins. These unusual spectra are different from those observed in ordinary solids or viscous liquids, but similar to those obtained in the case of diffusion in a cage, as discussed in the previous section. Above a critical temperature the spectra are composed of a narrow line, as in solids, and a broad component of comparable intensity. Similar spectra have also been observed in other iron-containing biological systems, e.g. in whole cells, membranes and tissue cultures (Nowik, Cohen, Bauminger & Ofer, 1983). Spectra obtained by quasi-elastic incoherent neutron scattering from nuclei in bound geometries display similar shapes.

In the early studies the experimental results obtained from protein crystals were treated by a mathematical model, which assumes that the iron Mössbauer probe responds to fluctuations between only two different molecular conformational substates (Mayo *et al.*, 1981). This now appears to have been an oversimplification and in subsequent investigations a model involving diffusion in a cage has been used. The most accepted model today is one in which the dynamics appropriate to the above situations is described in terms of harmonic oscillators in Brownian motion (Nowik *et al.*, 1983; Knapp, Fischer & Parak, 1983). This theory predicts that the Mössbauer spectra can be considered as being composed of a narrow and a broad line, with a total intensity which is strongly temperature dependent. When the diffusion is anisotropic a generalised Goldanskii–Karyagin effect can be expected, even when the *f*-factor is isotropic.

### 5.2　The theoretical model

#### 5.2.1　Translational bound diffusion

In recent studies two groups (Nowik *et al.*, 1983; Knapp *et al.*, 1983) have shown that Mössbauer nuclei which are harmonically bound to a centre by a harmonic force $m\Omega^2 r$, damped by a frictional force $m\beta\,dr/dt$ and acted upon by random forces will perform Brownian motion. In the overdamped limit $(\beta \gg \Omega)$ this yields Mössbauer spectra (Nowik *et al.*, 1983) given by

$$I(\omega) = \int_{-\infty}^{+\infty} \exp\left\{ i(\omega - \omega_0)t - \frac{\Gamma}{2}|t| - k^2\langle x^2\rangle [1 - \exp(-\alpha|t|)] \right\} dt$$

(6.17)

This equation was used to calculate the absorption spectra as a function of the parameters $\alpha = \Omega^2/\beta = 1/\tau_c$ and $D = \alpha\langle x^2\rangle$ where $\tau_c$ is the relaxation time of the ensemble average of $\langle x\rangle$ and $\langle x^2\rangle = k_B T/m\Omega^2$. Examples of spectral shapes computed in this way are shown in Figure 6.12 for a fixed value of $D$ and a range of values of $\alpha$. From Equation (6.17) it can be seen that for very small values of $\alpha$, a Lorentzian spectrum of linewidth $\Gamma + 2k^2 D$, as expected for a freely diffusing particle, is obtained. For large $\alpha$ $(\alpha/\Gamma = \Omega^2/\beta\Gamma \gg 1)$

and, while still keeping within the overdamped limit, $I(\omega)$ gives a Lorentzian line of width $\Gamma$. At intermediate values of $\alpha$ the spectra have anomalous shapes which are shown in Figure 6.12. Equation (6.17) can be represented by a rapidly converging infinite sum of Lorentzian lines. By expanding the exponential in a power series, $I(\omega)$ is given (Parak *et al.*, 1982) by

$$I(\omega) = \exp\left(-k^2\langle x^2\rangle\right) \sum_{n=0}^{\infty} \frac{(k^2\langle x^2\rangle)^n}{n!} \frac{[(\Gamma/2)+n\alpha]/\pi}{[(\Gamma/2)+n\alpha]^2 + (\omega-\omega_0)^2}$$

(6.18)

Equation (6.18) may be approximated by a sum of only two Lorentzian lines, one with a narrow linewidth, corresponding to $n=0$, and one with a broad linewidth, corresponding to $n>0$. The width of the broad line $\Gamma_B$ will

Fig. 6.12. Computed Mössbauer spectra of damped harmonically bound particles in Brownian motion as a function of $\alpha$, the ratio between the harmonic and the frictional force constants. (Nowik *et al.*, 1983.)

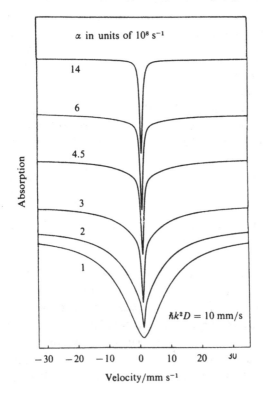

be given by the harmonic average

$$\frac{2\alpha}{\Gamma_B} = \frac{1}{\exp(k^2\langle x^2\rangle - 1)} \sum_{n=1}^{\infty} \frac{(k^2\langle x^2\rangle)^n}{n![(\Gamma/2\alpha)+n]} \tag{6.19}$$

Figure 6.13 shows $\Gamma_B/2\alpha$ as a function of $k^2\langle x^2\rangle$. In the two-line approximation $k^2\langle x^2\rangle$ can therefore be derived from the relative intensities of the two lines and $\alpha$ can then be obtained from Figure 6.13.

The model described above can be generalised to treat anisotropic diffusive motions which might be relevant in particular systems. For an anisotropic overdamped harmonic oscillator, with independent motions

Fig. 6.13. The theoretical width of the broad line (in units of $2\alpha$) as a function of $k^2\langle x^2\rangle$. The solid curved line corresponds to pure bound translational diffusion. The dashed line corresponds to a combination of bound translational and free rotational diffusion. The straight line corresponds to a linewidth of $2k^2D$, obtained for free diffusion. The insert shows the theoretical width of the narrow line in units of $\Gamma$ (the natural linewidth) as a function of $2k^2D/\Gamma$, for spheres of radius of 20 nm participating in bound translational and free rotational diffusive motions. (Ofer *et al.*, 1984.)

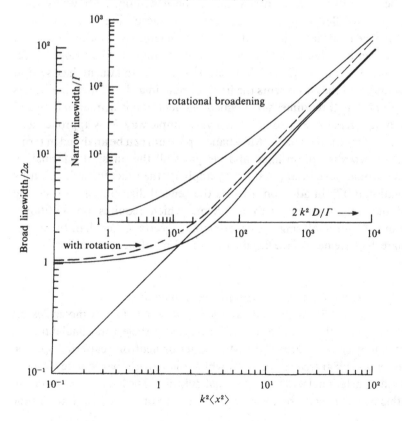

along the three axes, Equation (6.17) becomes

$$
I(\omega) = \frac{1}{2\pi} \int_{-\infty}^{+\infty} \left\{ \exp\left[ -i(\omega - \omega_0)t - \frac{\Gamma}{2}|t| - k_x^2 \langle x^2 \rangle (1 - e^{-\alpha_x|t|}) \right. \right.
$$

$$
\left. \left. - k_y^2 \langle y^2 \rangle (1 - e^{-\alpha_y|t|}) - k_z^2 \langle z^2 \rangle (1 - e^{-\alpha_z|t|}) \right] \right\} dt
$$

$$(6.20)$$

where $\alpha_x = \Omega_x^2/\beta_x$, $\alpha_y = \Omega_y^2/\beta_y$ and $\alpha_z = \Omega_z^2/\beta_z$. The lineshape obtained in a single protein crystal, or in oriented membranes, will then depend on the angle of incidence of the gamma rays. In polycrystalline proteins or unoriented membranes, in cases where there is an appreciable quadrupole interaction, a generalised Goldanskii–Karyagin effect might be expected, with the observed lineshape being a sum of different lines corresponding to the various transitions between the substates of the hyperfine energy level structure. Spectra calculated using Equation (6.20) for various values of $\alpha_x$ and $\alpha_y$ ($=\alpha_z$, the axial case) are shown in Figure 6.14. The asymmetry in the spectra is observable even when $\langle x^2 \rangle = \langle y^2 \rangle = \langle z^2 \rangle$.

The theoretical treatment which has been described above assumes harmonic oscillators with a single frequency according to the Einstein model. The treatment can be extended to a general distribution of normal frequencies and damping frequencies corresponding to the Debye model (Nowik, Bauminger, Cohen & Ofer, 1985). This treatment has several advantages in that it presents the final formula in a closed form and treats the problem in three dimensions, which enables the anisotropic binding and damping forces to be introduced in a very simple way. This approach also presents a means by which a Mössbauer spectrum can be analysed in terms of only two absorption lines, and yet gives all the physical information which would have been obtained by applying the exact formula given by Equation (6.17); in addition it treats the general distribution of restoring and damping frequencies in a scheme which requires no additional parameters for analysing the experimental spectra, other than the mean-square displacement of the fast fluctuations. For full details see Nowik *et al.* (1985).

### 5.2.2    Bound translational and free rotational diffusion

If the diffusing bound particle is composed of many molecules, as, for example, in the case of a rigid macroscopic particle, rotational diffusion may influence the shape of the Mössbauer or neutron scattering spectra. Ofer *et al.* (1984) treat the case in which the particle participates in both bound translational and free rotational diffusion. The Mössbauer spectrum in this case can again be represented by a sum of Lorentzian lines. In the simple case of a spherical homogeneous particle the spectrum is completely characterised by three parameters: the particle radius $r$, the mean-square displacement $\langle x^2 \rangle$ and the diffusion constant $D$, where $\alpha = D/\langle x^2 \rangle$ and the rotational diffusion constant is equal to $(\frac{3}{4})D/r^2$. For $k^2 \langle x^2 \rangle > 5$ the spectrum approximates to a single broadened Lorentzian line. In general, the effective linewidth $\Gamma_B$ depends on the size of the particle, although for $kr > 3$, $\Gamma_B = 2.542\, k^2 D$ and is independent of $r$. Thus for $^{57}$Fe, for which $k = 73\ \text{nm}^{-1}$, the effect of rotation is to increase the translational diffusion linewidth by only about 27 % (see Figure 6.13).

In the general case of a sphere performing bound translational and free rotational diffusion, the spectrum consists of a narrow line, the $n = 0$ line of Equation 6.18 broadened by rotation, and a broad line corresponding to

Fig. 6.14. Theoretical spectra of $^{57}$Fe under conditions of anisotropic bounded diffusion. (Bauminger *et al.*, 1983.)

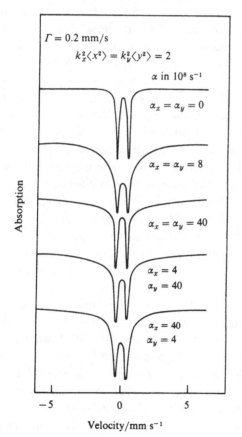

$\Gamma = 0.2$ mm/s
$k_x^2 \langle x^2 \rangle = k_y^2 \langle y^2 \rangle = 2$

$\alpha$ in $10^8$ s$^{-1}$

$\alpha_x = \alpha_y = 0$

$\alpha_x = \alpha_y = 8$

$\alpha_x = \alpha_y = 40$

$\alpha_x = 4$
$\alpha_y = 40$

$\alpha_x = 40$
$\alpha_y = 4$

Absorption

−5      0      5

Velocity/mm s$^{-1}$

$n > 1$, which is also broadened by rotation. The effective widths of these two lines have been calculated by harmonic averages (Ofer et al., 1984). Figure 6.13 also shows the width of the central narrow line in units of $\Gamma$, the natural linewidth, as a function of $2k^2D/\Gamma$.

### 5.3    Bounded diffusion in proteins

A sharp decrease in the recoil-free fraction $f$ around 210 K has been observed in solutions of nitrogenase, pea ferrodoxin, spermwhale myoglobin, human haemoglobin and a model iron–sulphur protein (Frolov, Likhtenstein & Goldanskii, 1975). This decrease in $f$ was only observed for samples above a critical relative moisture. Dwivedi, Pedersen & Debrunner (1979) have studied the temperature dependence of $f$ for deoxymyoglobin, oxymyoglobin, carboxymyoglobin and rubredoxin in frozen solutions. For temperatures above 180 K, $f$ dropped rapidly with temperature and differences due to the solvent matrix became apparent. Keller and Debrunner (1980) also found a sharp decrease in the $f$-factor in $^{57}$Fe enriched oxymyoglobin aqueous solution above 180 K. Above 240 K a diffusional line broadening, together with a decrease in quadrupole splitting due to random reorientation of the electric field gradient, were observed. Above 260 K the $f$-factor was so small that no Mössbauer spectrum could be obtained. Parak et al. (1981) measured the Mössbauer spectra of metmyoglobin crystals and again observed a sharp change in the temperature dependence of the $f$-factor at 210 K. These results were interpreted as being due to the appearance of a new motional degree of freedom, in addition to the usual vibrations present in any solid. The results were analysed assuming that the metmyoglobin molecule has only two different conformational substates with a certain energy difference between them. Fluctuations between these substates occur with temperature-dependent frequencies and in transitions from one substate to the other a potential barrier has to be surmounted. According to this analysis the iron moves together with part of the molecule over a distance of approximately 0.02 nm during the fluctuation and in each substate it vibrates with a frequency of about $10^{13}$ Hz which is similar to that in a solid.

Cohen et al. (1981) obtained Mössbauer spectra from crystals of the iron storage protein ferritin, and observed that, above 260 K, in addition to the normal quadrupole split doublet, the spectrum contained wide lines extending to velocities of $\pm 20$ mm/s. Mayo et al. (1981) observed similar spectra in carboxyhaemoglobin crystals which were enriched with $^{57}$Fe in the alpha chains. Similar spectra have also been observed in deoxymyoglobin crystals (Parak et al., 1982; Bauminger et al., 1983), in metmyoglobin crystals (Bauminger et al., 1983) and in other biological systems, and some characteristic spectra are shown in Figure 6.15. These spectra were analysed within the bounded diffusion model described above, as a function of $\alpha$, the ratio between the harmonic and damping force constants, and $\langle x^2 \rangle$, the mean-square displacement.

Some of the results from the measurements on deoxymyoglobin are shown in Figure 6.16, which displays the values of $-\ln f$ as a function of temperature, where $f$, the recoil-free fraction, is obtained from the area of

Fig. 6.15. Mössbauer absorption spectra in various biological systems containing $^{57}$Fe. The solid curves are theoretical least square fits. The parameters $D$ (in units of $10^{-14}$ m$^2$/s) and $\alpha$ (in units of $10^8$ s$^{-1}$) are respectively (a) 4, 2.5, (b) 40, 10, (c) 26, 5, and (d) 87, 13. (Nowik *et al.*, 1983.)

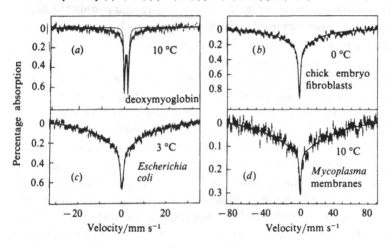

Fig. 6.16. Values of $-\ln f$ and $\langle x^2 \rangle$ for metmyoglobin corresponding to: (a) the area of the narrow subspectrum and (b) the total spectral area. (Bauminger *et al.*, 1983.)

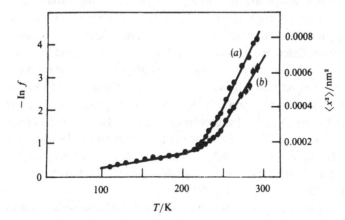

the absorption spectrum and is normalised to a value of one at 0 K. Figure 6.16(*a*) corresponds to the area of the narrow subspectrum, while Figure 6.16(*b*) corresponds to the total spectrum (narrow and wide components). Below a critical temperature $T_c = 220$ K only the narrow component was observed, whereas above 220 K the spectra were composed of narrow and wide components. Above 220 K the change of $-\ln f$ as a function of temperature is much steeper than below 220 K. From the values of $-\ln f$, the mean-square displacements $\langle x^2 \rangle$ were derived using the relation $f = \exp(-k^2 \langle x^2 \rangle)$. The values of $\langle x^2 \rangle$ are given on the right-hand side of Figure 6.16. The results obtained for metmyoglobin crystals were only slightly different from those for deoxymyoglobin. In ferritin a very sharp phase change was observed at $T_c = 260$ K. Below 260 K only the narrow subspectrum was observed, whereas above 260 K the spectra were composed of narrow and wide components, with values of $-\ln f$ and $\langle x^2 \rangle$ quite similar to those of myoglobin.

The values of $\langle x^2 \rangle$ below the critical temperature, derived from the areas of the spectra, are associated with the local thermal motions of the individual iron atoms relative to their neighbours. The total mean square displacements $\langle x^2 \rangle$ of the iron atoms above $T_c$, derived from the areas of the narrow subspectra, were resolved into three components associated with different modes of thermal motions: $\langle x^2 \rangle_{loc}$, $\langle x^2 \rangle_{sc}$ and $\langle x^2 \rangle_{fc}$ corresponding to local, slow collective and fast collective modes, respectively. This is shown in Figure 6.17 for metmyoglobin and ferritin. The value of $\langle x^2 \rangle_{loc}$ is obtained by a linear extrapolation from the values of $\langle x^2 \rangle$ below the critical temperature. The difference between $\langle x^2 \rangle$ and $\langle x^2 \rangle_{loc}$ gives the mean-square collective displacement $\langle x^2 \rangle_c$ associated with large-scale motions of parts of the surrounding protein and is resolved into the two parts: $\langle x^2 \rangle_{sc}$ and $\langle x^2 \rangle_{fc}$. The values of $\langle x^2 \rangle_{sc}$ correspond to slow collective motions which are responsible for the observed broad lines. These motions are characterised by a value of $2.5 \times 10^8$ Hz for $\alpha$, corresponding to a relaxation time of $4 \times 10^{-9}$ s. The values of $\langle x^2 \rangle_{sc}$ were obtained from the difference between the values of $\langle x^2 \rangle$ obtained from the area of the narrow subspectrum and the values of $\langle x^2 \rangle$ obtained from area of the total spectrum. The values of $\langle x^2 \rangle_{fc}$ were derived from the difference between $\langle x^2 \rangle_c$ and $\langle x^2 \rangle_{sc}$ and correspond to motions characterised by large values of $\alpha$, greater than $2 \times 10^9$ Hz. The broad lines associated with these motions are so broad that they cannot be observed experimentally. Parak *et al.* (1982) also observed a broadening of the narrow lines and attributed it to quasi-free diffusion, with $D$ having a value between $0.69 \times 10^{-17}$ m²/s and $9 \times 10^{-16}$ m²/s.

It is surprising that the parameters characterising the large-scale motions of the parts of the protein which produce the motion of the iron atoms, $\langle x^2 \rangle_{sc}$ and $\langle x^2 \rangle_{fc}$, are almost identical in myoglobin and ferritin. The bonding of iron in myoglobin is very different from that in ferritin; the iron is part of the haem group in myoglobin, whereas in ferritin the iron is in an essentially inorganic core with an approximate composition of FeOOH and a diameter of approximately 7 nm, which is located inside a protein shell. From a comparison between the results obtained in myoglobin and ferritin crystals the following conclusions can be drawn. The phase transitions taking place in iron-containing proteins are closely correlated with the state of the water in the proteins. In myoglobin the transition is gradual and corresponds to the gradual thawing of water absorbed in the protein. The sharp transition observed in ferritin very probably correlates with the melting of the water in the inner cavity of the protein. The temperature dependence of $\langle x^2 \rangle_{loc}$ is much steeper in myoglobin than in ferritin. This is consistent with the assumption that the bonding between the iron atoms in ferritin is more rigid than the bonding of the iron atom to the

Fig. 6.17. Values of total $\langle x^2 \rangle$, $\langle x^2 \rangle_{loc}$, $\langle x^2 \rangle_{sc}$, and $\langle x^2 \rangle_{fc}$ in metmyoglobin and in ferritin. (Ofer *et al.*, 1983.)

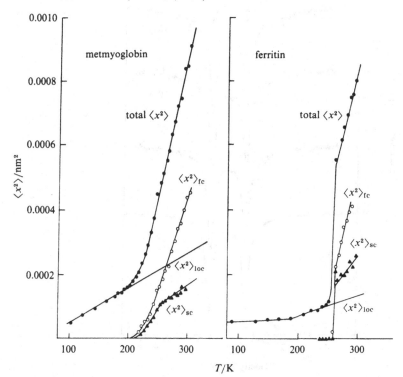

haem group of myoglobin. The similarity in $\langle x^2 \rangle_c$ indicates that the iron core of ferritin is composed of subaggregates which are not rigidly connected to each other and that each of the subaggregates is strongly bound to a protein subunit and is driven by the dynamics of that subunit. An appealing model for the collective motion in the two proteins, which might explain the similarity between them, is provided by the picture of hinge motion of their alpha helix structures which is strongly overdamped in the outer viscous glycerol-like layer of the protein.

In a detailed study of deoxymyoglobin crystals, an asymmetry in the spectra measured above $-25\,^\circ\text{C}$ has been observed (Bauminger *et al.*, 1983). The asymmetry in the spectral shapes shown in Figure 6.18 must be

Fig. 6.18. Mössbauer spectra of $^{57}$Fe in deoxymyoglobin crystals at various temperatures. (Bauminger *et al.*, 1983.)

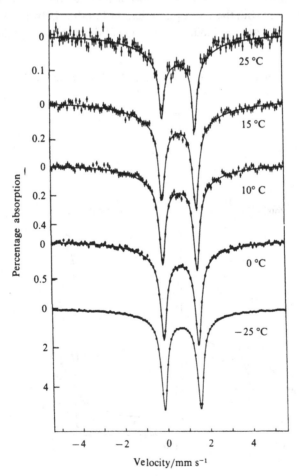

associated with anisotropic motion. The spectra have been analysed assuming isotropic damping frequencies (corresponding to $\beta_x = \beta_y = \beta_z$ in Equation (6.20)) and anisotropy in the mean-square displacements. In this analysis either a positive or negative quadrupole coupling constant $eQV_{zz}$ has been assumed, yielding the temperature dependences of $\langle x_\parallel^2 \rangle$ and $\langle x_\perp^2 \rangle$ which are shown in Figure 6.19. In both cases, the values for $\beta$ and $D$ are the same. In the temperature region studied $D$ is linear in temperature, and assuming $eQV_{zz}$ to be positive, values of 0.000 40 and 0.000 16 nm$^{-2}$ are obtained for $\langle x_\parallel^2 \rangle$ and $\langle x_\perp^2 \rangle$ at 20 °C.

In magnetotactic bacteria very broad Mössbauer spectra ($\Gamma \sim 100$ mm/s) have been observed above freezing temperatures (Ofer *et al.*, 1984). These spectra are due to the diffusive motion of the magnetosomes (enveloped

Fig. 6.19. Anisotropic mean square displacements and diffusion constant of $^{57}$Fe in deoxymyoglobin crystals. (Bauminger *et al.*, 1983.)

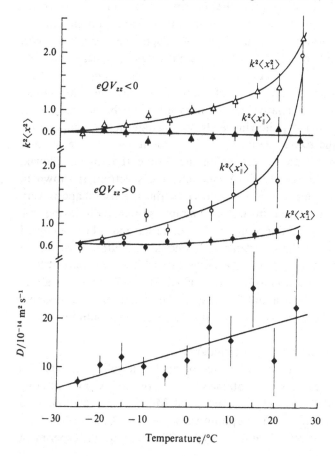

$Fe_3O_4$ particles) within the cytoplasmic fluid of the bacteria. The spectra were analysed within the extended bounded diffusion model discussed above, which provided information on the rotational and translational motions of the magnetosomes, as well as on the viscosity of the medium surrounding the magnetic particles. The results show that $\langle\theta^2\rangle^{\frac{1}{2}} < 1.5°$ and $\langle x^2\rangle^{\frac{1}{2}} < 0.84$ nm for the rotational and translational motions respectively. The effective viscosity coefficient is about 15 times greater than that of water at 295 K and increases with decreasing temperature.

### 5.4    Bounded diffusion in polymers

Polymer networks are complex structures with anomalous dynamic properties. Measurements on polymers based on vinyl ferrocene showed a change in the temperature dependence of the recoil-free fraction between 80 and 180 K, which was independent of changes in the polymer backbone (Litterst, Lerf, Nuyken & Alcala, 1982). In the same temperature region changes in the linewidths of the absorption lines were also observed. Both phenomena were explained by the thermally activated population of additional conformational substates. A copolymer swollen with dimethylformamide revealed a further sharp decrease in $f$ near to 190 K. This decrease was interpreted in terms of a glass transition, where cooperative motions of large parts of the backbone become important.

Spectra composed of relatively sharp lines and a very broad component of comparable intensity, showing a drastic change in the temperature dependence of the recoil-free fraction above about 180 K, have recently been observed in iron-containing Nafion membranes (Bauminger, Nowik, Ofer & Virguin-Heitner, 1985). A characteristic spectrum is shown in Figure 6.20. The temperature dependence of the spectral shape is very similar to that observed in iron-containing proteins displaying the phenomena of bounded diffusion. These spectra have also been analysed within the model of the overdamped harmonic oscillator in Brownian motion described in Section 5.2. The coexistence of broad and narrow lines is observable between about 180 and 240 K. At 240 K, $\langle x^2\rangle_{sc}$ is about $0.0004$ nm$^2$ and $\langle x^2\rangle_{fc}$ is about $0.0008$ nm$^2$. These motions are probably associated with the dynamics of the side chains of the membrane.

### 6    Comparisons with other techniques

Information on the dynamics of nuclei in solids, liquids and biological matter can also be obtained by neutron and X-ray scattering experiments, and by Rayleigh scattering of Mössbauer radiation. In a scattering experiment an incident photon of wavevector $k$ and energy $\hbar\omega$ is scattered with wavevector $k'$ and energy $\hbar\omega'$. Thus a scattering experiment has the advantage that the cross-section for scattering is measured as a function of two independent parameters, related to the momentum and energy transfer to the system. In a Mössbauer study the absorption is only measured as a function of energy. This limitation is compensated by the high energy resolution of Mössbauer spectroscopy.

Mössbauer (1983) has reviewed the information obtained from Mössbauer spectroscopy in comparison with that obtained by other techniques, and has shown the similarities and differences in the expressions for the cross-sections for the various processes associated with each of the techniques mentioned above. Since the Mössbauer absorption process is incoherent and only yields information about the individual motion of a particle, the comparison is made with similar incoherent processes. Following a double Fourier transform the experimental cross-sections as a function of momentum and frequency yield the van Hove correlation function. This function contains the probability distribution of the absorbing or scattering atoms as a function of space and time, which is the main goal of all these studies. Information on short-term behaviour is contained in measurements at high frequencies, while measurements at low frequencies, corresponding to small energy transfers, give information on the long-term motional behaviour. Similarly, small momentum transfers sample large regions of space, while the motional behaviour over short distances reveals itself in measurements at large momentum transfers, corresponding to large changes in the wavevector. These Fourier

Fig. 6.20. The Mössbauer spectrum of $^{57}$Fe in divalent iron in Nafion membranes at 230 K. The fitted lines show the decomposition of the spectrum into the elastic narrow doublet and the quasi-elastic broad line.

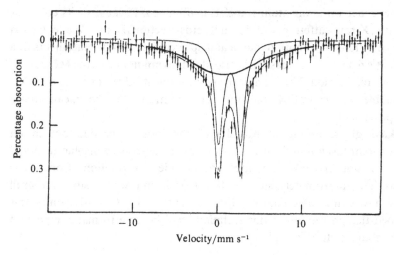

transforms represent strong correlations between time and frequency and momentum and space, and hence the relative advantages of the various techniques can be easily evaluated.

In elastic processes, which correspond to zero energy transfer, all the experimental techniques under consideration, i.e. neutron, X-ray, Rayleigh and Rayleigh–Mössbauer scattering, yield the momentum transfer dependence of the Debye–Waller or Mössbauer $f$-factors. These elastic processes only arise if the position coordinate of the particle is confined to a finite region of space, since free diffusion such as occurs in liquids does not give rise to an elastic process.

When inelastic processes are of interest, neutron scattering is the most valuable technique. It is the only technique which allows, in principle at least, the entire range of energy and momentum transfer to be scanned. Neutron scattering is very effective in studies of large energy transfers and thus is widely used to determine phonon dispersion relations in solids. Because of the limited resolution in neutron scattering experiments, the studied timescale of atomic motion is in the range between $10^{-12}$ and $10^{-10}$ s. Mössbauer spectroscopy, with its high energy resolution, has a sensitivity to atomic motions with timescales in the range between about $0.01 \tau_N$ and $10 \tau_N$, where $\tau_N$ is the mean lifetime of the Mössbauer excited state. For $^{57}$Fe this corresponds to around $10^{-9}$ to $10^{-6}$ s. As we have seen in the previous section, particle motions in proteins, in polymers and in various biological systems are associated with timescales of the order of $\tau_N$ for the $^{57}$Fe Mössbauer isotope.

X-ray scattering has good momentum transfer resolution, but very poor energy resolution and thus does not yield information on the dynamical behaviour of atoms. Only the static Debye–Waller factor, with its dependence on momentum transfer, is obtained. Because of its coherent nature, X-ray scattering contains in its cross-section (Debye–Waller factor) contributions from both dynamical displacements and displacements due to static disorder, whereas static disorder does not influence the Mössbauer recoil-free fraction. Thus comparison of the results of the two techniques is valuable in separating the two contributions to the mean-square displacements.

Rayleigh scattering of Mössbauer radiation is carried out in an incoherent fashion and thus measures the mean-square displacements of atomic motions only and does not include contributions from static disorder. Unfortunately, such studies suffer from low intensities and until now have contributed relatively little to the study of atomic dynamics. It is hoped that in the future this technique will be able to make important contributions to this field.

7    **Conclusions**

In the early days of the technique Mössbauer spectroscopy was strongly involved in the study of nuclear dynamics. For a while the interest in the study of dynamics through Mössbauer spectroscopy faded, mainly because this work involved accurate measurements of intensities or very small changes in line positions, and these are not easy to carry out. However, the more recent observations that phenomena of nuclear dynamics can be observed through the shape of the Mössbauer spectrum has brought a revival of interest in this field of research.

It has been found that Mössbauer spectroscopic studies in diffusing systems can give deeper insight into the microscopic mechanism of diffusion. In free diffusion a temperature-dependent broadened linewidth is observed. In liquids the viscosity coefficient as a function of temperature can be determined. In certain cases rotational diffusion may produce a change in the lineshape. In solids, information on the diffusion constant and the jump frequency and their temperature dependence, and on the energy barrier between conformational states, can be obtained. Anisotropic diffusion in polycrystalline samples will cause different broadening of the various hyperfine components in the spectrum and a deviation from a Lorentzian lineshape, resulting in a cusp-like profile. The deviation from a Lorentzian lineshape is difficult to observe and special techniques have to be used to detect it. In anisotropic diffusion in single crystals and in liquid crystals, broadened Lorentzian lines, with different widths at different orientations of the single crystal, are observed. From the angular dependence and temperature dependence of the recoil-free fraction and the linewidth, information can be obtained on the diffusion geometry, as well as on the diffusion constant, jump frequency and energy barrier.

Bounded diffusion in all the systems investigated (polymers, proteins, living cells, membranes, etc.) produces in certain temperature regions unusual Mössbauer lineshapes composed of a narrow elastic line and a broad quasi-elastic line. In cases where hyperfine structure is present, or in single crystals, the anisotropy in the dynamics of the absorbing nuclei, and the gradual change in the spectral shape as a function of the parameters associated with the dynamics, can be observed. Similar phenomena have also been observed when the Mössbauer nucleus itself is not diffusing, but is moving as a result of the diffusion of its neighbours. The shape of the spectra might also be affected by relaxation phenomena caused by the diffusive motion, which will usually cause motional narrowing at high diffusion rates (see Chapter 5). The new phenomena observed open up many areas for possible future research.

It therefore seems that the study of nuclear dynamics by Mössbauer spectroscopy is still at an early stage. Because of the wide interest in the dynamics of biological and other macromolecular systems and the potential information the Mössbauer technique can supply, this field is certainly likely to develop in the near future. Many systems appear to exhibit broad quasi-elastic Mössbauer absorption lines. In the past these were not noticed because most work with $^{57}$Fe involved spectra measured within a velocity range of $\pm 10$ mm/s, and with statistics which were inadequate for the observation of the hidden broad line. The recent advances in the construction of gamma-ray detection systems capable of very high count rates of up to $5 \times 10^5$ counts per second have enabled good $^{57}$Fe Mössbauer spectra to be obtained, even when the absorption intensity is only 0.1% or less.

Clearly it should be possible to carry out nuclear dynamics measurements of the type discussed in this chapter with isotopes other than $^{57}$Fe. The isotopes $^{119}$Sn, $^{151}$Eu and $^{161}$Dy would appear to be good candidates for work of this type.

### Acknowledgement

This work was supported in part by the Fund for Basic Research, administered by the Israel Academy of Sciences and Humanities.

# References

## Chapter 1

Cohen, R. L., ed. (1976 and 1980). *Applications of Mössbauer Spectroscopy*, Vols. I and II, Academic Press, New York.

Gibb, T. C. (1977). *Principles of Mössbauer Spectroscopy*, Chapman and Hall, London.

Gonser, U., ed. (1975). *Topics in Applied Physics* Vol. V: *Mössbauer Spectroscopy*, Springer-Verlag, Berlin.

Gonser, U., ed. (1981). *Mössbauer Spectroscopy II: The Exotic Side of the Method*, Springer-Verlag, Berlin.

Greenwood, N. N. & Gibb, T. C. (1971). *Mössbauer Spectroscopy*, Chapman and Hall, London.

Gruverman, I. J., ed. (1965–74). *Mössbauer Effect Methodology*, Vols. I–IX, Plenum Press, New York.

Long, G. J., ed. (1984). *Mössbauer Spectroscopy Applied to Inorganic Chemistry*, Vol. I, Plenum Press, New York.

May, L. (1971). *An Introduction to Mössbauer Spectroscopy*, Plenum Press, New York.

Mössbauer, R. L. (1958). *Z. Physik*, **151**, 124.

Thosar, B. V., Srivastava, J. K., Iyengar, P. K. & Bhargava, S. C., eds. (1983). *Advances in Mössbauer Spectroscopy, Applications to Physics, Chemistry and Biology*, Elsevier, Amsterdam and New York.

Wertheim, G. K. (1964). *Mössbauer Effect Principles and Applications*, Academic Press, New York.

## Chapter 2

Adams, I., Thomas, J. M., Bancroft, G. M., Butler, K. D. & Barber, M. (1972). *J. Chem. Soc., Chem. Commun.*, p. 751.

Appleton, T. G., Clark, H. C. & Manzer, L. E. (1973). *Coord. Chem. Rev.*, **10**, 335.

Armstrong, D. A., Fortune, R., Perkins, P. G., Dickinson, R. J. & Parish, R. V. (1976). *Inorg. Chim. Acta*, **17**, 73.

Baldy, A., Limouzin, Y., Maire, J. C. & Mondou, J. C. (1976). *J. Microsc. Spectrosc. Electron*, **1**, 469.

Baltrunas, D. I., Ionov, S. P., Aleksandrov, A. Yu. & Makarov, E. F. (1973). *Chem. Phys. Lett.*, **20**, 555.

Bancroft, G. M., Mays, M. J. & Prater, B. E. (1970). *J. Chem. Soc. (A)*, p. 956.

Bancroft, G. M., Mays, M. J., Prater, B. E. & Stefanini, F. P. (1970). *J. Chem. Soc. (A)*, p. 2146.

## 262    References

Bancroft, G. M., Garrod, R. E. B. & Maddock, A. G. (1971). *J. Chem. Soc. (A)*, p. 3165.

Bancroft, G. M. & Platt, R. H. (1972). *Adv. Inorg. Chem. Radiochem.*, **15**, 59.

Bancroft, G. M., Butler, K. D. & Libbey, E. T. (1972). *J. Chem. Soc., Dalton Trans.*, p. 2643.

Bancroft, G. M. & Libbey, E. T. (1973). *J. Chem. Soc., Dalton Trans.*, p. 2103.

Bancroft, G. M. & Butler, K. D. (1974). *J. Amer. Chem. Soc.*, **96**, 7208.

Bancroft, G. M. & Sham, T. K. (1974). *Canad. J. Chem.*, **52**, 1361.

Barber, M., Swift, P., Cunningham, D. & Frazer, M. J. (1970). *Chem. Commun.*, p. 1338.

Barbieri, R., Pellerito, L., Bertazzi, N. & Stocco, G. C. (1973). *Inorg. Chim. Acta*, **11**, 74.

Bartunik, H. D., Holzapfel, W. F. & Mössbauer, R. L. (1970). *Phys. Lett.*, **33A**, 469.

Bent, H. A. (1961). *Chem. Rev.*, **61**, 1.

Berrett, R. R. & Fitzsimmons, B. W. (1967). *J. Chem. Soc. (A)*, p. 525.

Berrett, R. R., Fitzsimmons, B. W. & Owusu, A. (1968). *J. Chem. Soc. (A)*, p. 1575.

Bertazzi, N., Gibb, T. C. & Greenwood, N. N. (1976). *J. Chem. Soc., Dalton Trans.*, p. 1153.

Birchall, T. & Della Valle, B. (1971). *Canad. J. Chem.*, **49**, 2808.

Birchall, T., Della Valle, B., Martineau, E. & Milne, J. B. (1971). *J. Chem. Soc. (A)*, p. 1855.

Birchall, T., Ballard, J. G., Milne, J. B. & Moffatt, W. D. (1974). *Canad. J. Chem.*, **52**, 2375.

Brill, T. B., Parris, G. E., Long, G. J. & Bowen, L. H. (1973). *Inorg. Chem.*, **12**, 1888.

Bukshpan, S., Goldstein, C. & Sonnino, T. (1968). *J. Chem. Phys.*, **49**, 6477.

Cheyne, B. M., Johnston, J. J. & Jones, C. H. W. (1972). *Chem. Phys. Lett.*, **14**, 545.

Clark, M. G., Maddock, A. G. & Platt, R. H. (1972). *J. Chem. Soc., Dalton Trans.*, p. 281.

Clark, R. J. H., Maresca, L. & Smith, P. J. (1970). *J. Chem. Soc., Dalton Trans.*, p. 2687.

Dailey, B. P. & Townes, C. H. (1955). *J. Chem. Phys.*, **23**, 118.

Dale, B. W., Dickinson, R. J. & Parish, R. V. (1979). *Chem. Phys. Lett.*, **64**, 375.

Das, T. P. & Hahn, E. L. (1958). *Solid State Phys. Suppl.*, **1**, 1.

Davies, A. G., Milledge, H. J., Puxley, D. C. & Smith, P. J. (1970). *J. Chem. Soc. (A)*, p. 2862.

Dickinson, R. J., Parish, R. V. & Dale, B. W. (1980). *J. Chem. Soc., Dalton Trans.*, p. 895.

Dickinson, R. J., Parish, R. V., Rowbotham, P. J., Manning, A. R. & Hackett, P. (1975). *J. Chem. Soc., Dalton Trans.*, p. 424.

Dirac, P. A. M. (1928). *Proc. Roy. Soc.* (London), **117A**, 610 and **118A**, 351.

Donaldson, J. D., Filmore, E. J. & Tricker, M. J. (1971). *J. Chem. Soc. (A)*, p. 1109.

Donaldson, J. D., Southern, J. T. & Tricker, M. J. (1972). *J. Chem. Soc., Dalton Trans.*, p. 2637.

Ehrlich, B. S. & Kaplan, M. (1971). *J. Chem. Phys.*, **54**, 612.

Fischer, G. F. (1977). *The Hartree–Fock Method for Atoms*, Wiley, New York.

Freeman, A. J. & Ellis, D. E. (1978). In: *Mössbauer Isomer Shifts*, eds. F. E. Wagner & G. K. Shenoy, Chap. 4, North Holland, Amsterdam.

Fryer, C. W. & Smith, J. A. S. (1970). *J. Chem. Soc. (A)*, p. 1029.

Furlani, C., Mattogno, G., Polzonetti, G., Barbieri, R., Rivarola, E. & Silvestri, A. (1981). *Inorg. Chim. Acta*, **52**, 23.

Gibb, T. C. (1968). *J. Chem. Soc. (A)*, p. 1439.

Gibb, T. C., Goodman, B. A. & Greenwood, N. N. (1970). *Chem. Commun.*, p. 774.

Golding, R. M. (1976). *Mol. Phys.*, **12**, 13.

Greenwood, N. N., Perkins, P. G. & Wall, D. A. (1967). *Symp. Farad. Soc.*, **1**, 51.

Grutsch, P. A., Zeller, M. V. & Fehlner, T. P. (1973). *Inorg. Chem.*, **12**, 1431.

Gütlich, P., Link, R. & Trautwein, A. (1978). *Mössbauer Spectroscopy and Transition Metal Chemistry*, p. 56, Springer-Verlag, Berlin.

Harris, R. K. & Mann, B. E. (1978). *NMR and the Periodic Table*, Academic Press, New York.

Holmes, J. R. & Kaesz, H. D. (1961). *J. Amer. Chem. Soc.*, **83**, 3903.

Hossain, M. B., Lefferts, J. L., Molloy, K. C., van der Helm, D. & Zuckerman, J. J. (1979). *Inorg. Chim. Acta*, **36**, L409.

Ingalls, R. (1964). *Phys. Rev.*, **133A**, 787.

Ingalls, R., van der Woude, F. & Sawatzky, G. A. (1978). In: *Mössbauer Isomer Shifts*, eds. F. E. Wagner & G. K. Shenoy, p. 357, North Holland, Amsterdam.

Jones, C. H. W. (1975). *J. Chem. Phys.*, **62**, 4343.

Jones, C. H. W., Schultz, R., McWhinnie, W. R. & Dance, N. S. (1976). *Canad. J. Chem.*, **54**, 3234.

Jones, C. H. W. & Dombsky, M. (1981). *Can. J. Chem.*, **59**, 1585.

Kaindl, G., Potzel, W., Wagner, F. E., Zahn, U. & Mössbauer, R. L. (1969). *Z. Physik*, **226**, 103.

Kerler, W. & Neuwirth, W. (1962). *Z. Physik.*, **167**, 176.

Langouche, G., Triplett, B. B., Dixon, N. D., Mahmud, Y. & Hanna, S. S. (1977). *Phys. Rev.*, **B15**, 2504.

Long, G. J., Stevens, J. G., Tullbane, R. J. & Bowen, L. H. (1970). *J. Amer. Chem. Soc.*, **92**, 4230.

Marks, A. P., Drago, R. S., Herber, R. H. & Potasek, M. J. (1976). *Inorg. Chem.*, **15**, 259.

Mays, M. J. & Sears, P. L. (1974). *J. Chem. Soc., Dalton Trans.*, p. 2254.

McAuliffe, C. A., Niven, I. E. & Parish, R. V. (1977*a*). *J. Chem. Soc., Dalton Trans.*, p. 1670.

McAuliffe, C. A., Niven, I. E. & Parish, R. V. (1977*b*). *J. Chem. Soc., Dalton Trans.*, p. 1901.

McFarlane, W. (1967). *J. Chem. Soc. (A)*, p. 528.

McGrady, M. M. & Tobias, R. S. (1965). *J. Amer. Chem. Soc.*, **87**, 1909.

Micklitz, H. & Barrett, P. H. (1972). *Phys. Rev.*, **B5**, 1704.

Miller, G. A. & Schlemper, E. O. (1973). *Inorg. Chem.*, **12**, 677.

Mitchell, T. N. (1973). *J. Organomet. Chem.*, **59**, 189.

Parish, R. V. & Platt, R. H. (1969). *J. Chem. Soc. (A)*, p. 2145.

Parish, R. V. & Platt, R. H. (1970). *Inorg. Chim. Acta*, **4**, 65.

Parish, R. V. & Johnson, C. E. (1971). *J. Chem. Soc. (A)*, p. 1906.

Parish, R. V. & Rowbotham, P. J. (1971). *Chem. Phys. Lett.*, **11**, 137.

Parish, R. V. (1972). *Progr. Inorg. Chem.*, **15**, 101.

Parish, R. V. (1982*c*). *Coord. Chem. Rev.*, **42**, 1.

Parish, R. V. (1982*a*). *Gold Bulletin*, **15**, 51.

Parish, R. V. (1982*b*). *Mössbauer Effect. Ref. Data J.*, **5**, 196.

Parish, R. V. (1984*a*). In: *Mössbauer Spectroscopy Applied to Inorganic Chemistry*, Vol. I, ed. G. J. Long, Chap. 18, Plenum Press, New York.

Parish, R. V. (1984*b*). *Ibid.*, Chap. 19.

Parr, R. G. & Pearson, R. G. (1983). *J. Amer. Chem. Soc.*, **105**, 7512.

Pasternak, M. & Barrett, P. H. (1981). *Chem. Phys. Lett.*, **78**, 174.

Pasternak, M. & Bukshpan, S. (1967). *Phys. Rev.*, **163**, 297.

Pasternak, M., Simopoulos, A. & Hazony, Y. (1965). *Phys. Rev.*, **140A**, 1892.

Pasternak, M. & Sonnino, T. (1968). *J. Chem. Phys.*, **48**, 1997.

Perlow, G. J. & Perlow, M. R. (1966). *J. Chem. Phys.*, **45**, 2193.

Pitzer, K. S. (1979). *Acc. Chem. Res.*, **12**, 271.

Potzel, W., Wagner, F. E., Zahn, U., Mössbauer, R. L. & Danon, J. (1970). *Z. Physik.*, **240**, 306.

Powell, R. E. (1968). *J. Chem. Educ.*, **45**, 558.

Prosser, H., Wagner, F. E., Wortmann, G., Kalvius, G. M. & Wappling, R. (1975). *Hyperfine Interactions*, **1**, 25.

Pyyko, P. & Desclaux, J. P. (1979). *Acc. Chem. Res.*, **12**, 271.

Reschke, R., Trautwein, A. & Desclaux, J. P. (1977). *J. Phys. Chem. Solids*, **38**, 837.

Ruby, S. L., Kalvius, G. M., Beard, G. B. & Snyder, R. E. (1967). *Phys. Rev.*, **159**, 239.
Ruby, S. L. & Shenoy, G. K. (1978). In: *Mössbauer Isomer Shifts*, eds. F. E. Wagner & G. K. Shenoy, Chap. 9b, North Holland, Amsterdam.
Schlemper, E. O. (1967). *Inorg. Chem.*, **6**, 2012.
Sham, T. K. & Bancroft, G. M. (1975). *Inorg. Chem.*, **14**, 2281.
Sham, T. K., Watson, R. E. & Perlman, M. L. (1981). *Adv. Chem. Ser.*, **194**, 39.
Smith, P. J. (1970). *Organomet. Chem. Rev.*, **A5**, 373.
Sternheimer, R. M. (1956). *Phys. Rev.*, **102**, 73, and references therein.
Swartz, W. E., Watts, P. H., Lippincott, E. R., Watts, J. C. & Huheey, J. E. (1972). *Inorg. Chem.*, **11**, 2632.
Trautwein, A., Freeman, A. J. & Desclaux, J. P. (1976). *Phys. Rev.* **B13**, 1884.
Wagner, F. E., Potzel, W., Wagner, U. & Schmidtke, H. H. (1974). *Chem. Phys.*, **4**, 284.
Wagner, F. E. & Wagner, U. (1978). In: *Mössbauer Isomer Shifts*, eds. F. E. Wagner & G. K. Shenoy, Chap. 8a, North Holland, Amsterdam.
Walker, L. R., Wertheim, G. K. & Jaccarino, V. (1961). *Phys. Rev. Lett.* **6**, 98.
Williams, A. F., Jones, G. C. H. & Maddock, A. G. (1975). *J. Chem. Soc., Dalton Trans.*, p. 1952.

## Chapter 3

Asch, L., Shenoy, G. K., Friedt, J. M., Adloff, J. P. & Kleinberger, R. R. (1975a). *J. Chem. Phys.*, **62**, 2335.
Asch, L., Shenoy, G. K., Friedt, J. M. & Adloff, J. P. (1975b). *J. Chem. Soc. Dalton Trans.*, p. 1235.
Bargeron, C. B., Avinor, M. & Drickamer, H. G. (1971). *Inorg. Chem.*, **10**, 1338.
Bates, A. R. & Stevens, K. W. H. (1969). *J. Phys. C.*, **2**, 1573.
Bates, A. R. (1970). *J. Phys. C.*, **3**, 1825.
Bénard, M. (1982). In: *Electron Distributions and The Chemical Bonds*, eds. P. Coppens & M. B. Hall, p. 221, Plenum Press, New York.
Benson, C. G., Long, G. J., Kolis, J. W. & Shriver, D. F. (1985). *J. Amer. Chem. Soc.*, **107**, 5297.
Bentley, R. B., Gerloch, M., Lewis, J. & Quested, P. N. (1971). *J. Chem. Soc. (A)*, p. 3751.
Berrett, R. R. & Fitzsimmons, B. W. (1967). *J. Chem. Soc. (A)*, p. 525.
Berrett, R. R., Fitzsimmons, B. W. & Owusu, A. (1968). *J. Chem. Soc. (A)*, p. 1575.
Bode, K., Gütlich, P. & Koppen, H. (1980). *Inorg. Chim. Acta*, **42**, 281.
Bonnette, A. K. & Allen, J. F. (1971). *Inorg. Chem.*, **10**, 1613.
Brown, D. B. & Wrobleski, J. T. (1980). In: *Mixed Valence Compounds*, ed. D. B. Brown, p. 243, Reidel, Boston.
Brunot (1974). *Chem. Phys. Lett.*, **29**, 368 and 371; *J. Chem. Phys.*, **61**, 2360.
Bryan, R. F., Greene, P. T., Newlands, M. J. & Field, D. S. (1970). *J. Chem. Soc. (A)*, p. 3068.
Buser, H. J. Schwarzenbach, D., Petter, W. & Ludi, A. (1977). *Inorg. Chem.*, **16**, 2704.
Cheetham, A. K., Cole, A. J. & Long, G. J. (1981). *Inorg. Chem.*, **20**, 2747.
Clarke, P. J. & Milledge, H. J. (1975). *Acta Cryst.*, **31B**, 1543, 1554.
Collins, R. L. (1965). *J. Chem. Phys.*, **42**, 1072.
Collins, R. L., Pettit, R. & Baker, W. A. (1966). *J. Inorg. Nucl. Chem.*, **28**, 1001.
Coppens, P., Li, L. & Zhu, N. J. (1983). *J. Amer. Chem. Soc.*, **105**, 6173.
Cotton, F. A. & Troup, J. M. (1974). *J. Amer. Chem. Soc.*, **96**, 4155.
Cowan, D. O., Collins, R. L. & Kaufman, F. J. (1971). *J. Phys. Chem.*, **75**, 2025.
Cramer, R. E., Higa, K. T., Pruskin, S. L. & Gilje, J. W. (1983). *J. Amer. Chem. Soc.*, **105**, 6749.
Dahl, L. F. & Rundle, R. E. (1957). *Chem. Phys.*, **27**, 323.
Dahl, L. F. & Blount, J. F. (1965). *Inorg. Chem.*, **4**, 1373.

Dale, B. W. (1971). *J. Phys. C*, **4**, 2705.

Dale, B. W. (1974). *Molec. Phys.*, **28**, 503.

Dedieu, A., Rohmer, M. M. & Veillard, A. (1982). *Adv. Quant. Chem.*, **16**, 43.

Decurtins, S., Gütlich, P., Köhler, C. P., Spiering, H. & Hauser, A. (1984). *Chem. Phys. Lett.*, **105**, 1.

Decurtins, S., Gütlich, P., Hasselbach, K. M., Hauser, A. & Spiering, H. (1985). *Inorg. Chem.*, **24**, 2174.

Desiderato, R. & Dobson, G. R. (1982). *J. Chem. Ed.*, **59**, 752.

Dezsi, I., Keszthelyi, L., Molnar, B. & Pocs, L. (1965). *Phys. Lett.*, **18**, 28.

Dezsi, I., Keszthelyi, L., Molnar, B., & Pocs, L. (1968). In: *Hyperfine Structure and Nuclear Radiations*, eds. E. Mathias & D. Shirley, p. 566, North Holland, Amsterdam.

Dolphin, D. H., Sams, J. R., Tsin, T. B. & Wong, K. L. (1978). *J. Amer. Chem. Soc.*, **100**, 1711.

Eismann, G. A. & Reiff, W. M. (1980). *Inorg. Chem. Acta*, **44**, L171.

Erickson, N. E. & Fairhall, A. W. (1965). *Inorg. Chem.*, **4**, 1320.

Figgis, B. N. & Wadley, L. G. B. (1972). *Aust. J. Chem.*, **25**, 233.

Finklea, S. L., Cathey, L. & Amma, E. L. (1976). *Acta Cryst.*, **32A**, 529.

Fitzsimmons, B. W. (1980). *J. Physique*, **41**, C1–33.

Fitzsimmons, B. W. & Hume, A. R. (1980). *J. Chem. Soc.*, *Dalton Trans.*, p. 180.

Fluck, E., Kerler, W. & Neuwirth, W. (1963). *Angew. Chem. Int. Ed. Engl.*, **2**, 277.

Franke, P. L., Haasnoot, J. G. & Zuur, A. P. (1982). *Inorg. Chim. Acta*, **59**, 5.

Ganguli, P., Gütlich, P., Müller, E. W. & Irler, W. (1981). *J. Chem. Soc.*, *Dalton Trans.*, p. 441.

Ganguli, P., Gütlich, P. & Muller, E. W. (1982). *Inorg. Chem.*, **21**, 3429.

Gibb, T. C. (1976). *J. Phys. C.*, **9**, 2627.

Gibb, T. C., Greatrex, R., Greenwood, N. N. & Thompson, D. T. (1967). *J. Chem. Soc. (A)*, p. 1663.

Goodenough, J. B. (1980). In: *Mixed Valence Compounds*, ed. D. B. Brown, p. 413, Reidel, Boston.

Grandjean, F. & Gerard, A. (1975). *Solid State Commun.*, **16**, 553.

Greenwood, N. N. & Gibb, T. C. (1971). *Mössbauer Spectroscopy*, pp. 66, 91, Chapman and Hall, London.

Gütlich, P. (1981). *Structure and Bonding*, **44**, 83.

Gütlich, P. (1984). In: *Mössbauer Spectroscopy Applied to Inorganic Chemistry*, Vol. I, ed G. J. Long, p. 287, Plenum Press, New York.

Hall, W., Kim, S., Zubieta, J., Walton, E. G. & Brown, D. B. (1977). *Inorg. Chem.*, **16**, 1884.

Harris, G. (1968). *Theor. Chim. Acta*, **10**, 119.

Heilmann, I. U., Olsen, N. B. & Olsen, J. S. (1977). *Physica Scripta*, **15**, 285.

Herber, R. H., Kingston, W. R. & Wertheim, G. K. (1963). *Inorg. Chem.*, **2**, 153.

Herren, F., Fischer, P., Ludi, A. & Hälg, W. (1980). *Inorg. Chem.*, **19**, 956.

Hieber, W. & Becker, E. (1930). *Ber.*, **63**, 1405.

Holladay, A., Leung, P. & Coppens, P. (1983). *Acta Cryst.*, **39A**, 377.

Hoy, G. R. (1984). In: *Mössbauer Spectroscopy Applied to Inorganic Chemistry*, Vol. I, ed. G. J. Long, p. 195, Plenum Press, New York.

Hudson, A. & Whitfield, H. J. (1967). *Inorg. Chem.*, **6**, 1120.

Ingalls, R. (1964). *Phys. Rev.*, **133A**, 787.

Johnson, B. F. G. (1976). *J. Chem. Soc.*, *Chem. Commun.*, p. 703.

Jones, E. R., Hendricks, M. E., Auel, T. & Amma, E. L. (1977). *J. Chem. Phys.*, **66**, 3252.

Kalvius, M., Zahn, U., Kienle, P. & Eicher, H. (1962). *Z. Naturforsch*, **17A**, 494.

Kim, P. H. (1960). *J. Phys. Soc. Japan*, **15**, 445.

Kirner, J. F., Dow, W. & Scheidt, W. R. (1976). *Inorg. Chem.*, **15**, 1686.

Klemm, L. & Klemm, W. (1935). *J. Prakt. Chem.*, **143**, 82.

266     References

König, E. & Madeja, K. (1967). *Inorg. Chem.*, **6**, 48.
König, E., Hufner, S., Steichele, E. & Madeja, K. (1967 and 1968). *Z. Naturforsch*, **22A**, 1543; **23A**, 632.
Lecomte, C., Chadwick, D. L., Coppens, P. & Stevens, E. D. (1983). *Inorg. Chem.*, **22**, 2982.
Lever, A. B. P. (1966). *J. Chem. Soc.*, p. 1821.
Litterst, F. J. & Amthauer, G. (1984). *Phys. Chem. Minerals*, **10**, 250.
Little, B. F. & Long, G. J. (1978). *Inorg. Chem.*, **17**, 3401.
Long, E. A. & Toettcher, F. C. (1946). *J. Chem. Phys.*, **8**, 504.
Long, G. J. (1968). Doctoral Thesis, Syracuse University.
Long, G. J., Whitney, D. L. & Kennedy, J. E. (1971). *Inorg. Chem.*, **10**, 1406.
Long, G. J. & Coffen, D. L. (1974). *Inorg. Chem.*, **13**, 270.
Long, G. J. & Clarke, P. J. (1978). *Inorg. Chem.*, **17**, 1394.
Long, G. J. (1984). In: *Mössbauer Spectroscopy Applied to Inorganic Chemistry*, Vol. I, ed. G. J. Long, p. 7, Plenum Press, New York.
Long, G. J., Russo, U. & Benson, C. G., unpublished results.
Maer, K., Beasley, M. L., Collins, R. L. & Milligan, W. O. (1968). *J. Amer. Chem. Soc.*, **90**, 3201.
Mitschler, A., Rees, B. & Lehmann, M. S. (1978). *J. Amer. Chem. Soc.*, **100**, 3390.
Montano, P. A. & Seehra, M. S. (1976). *Solid State Commun.*, **20**, 897.
Morrison, W. H. & Hendrickson, D. N. (1975). *Inorg. Chem.*, **14**, 2331.
Moss, T. H. & Robinson, A. B. (1968). *Inorg. Chem.*, **7**, 1692.
Müller, E. W., Ensling, J., Spiering, H. & Gütlich, P. (1983). *Inorg. Chem.*, **22**, 2074.
Narath, A. (1965). *Phys. Rev.* **139A**, 1221.
Nolet, D. A. & Burns, R. G. (1979). *Phys. Chem. Minerals*, **4**, 221.
Purcell, K. F. & Kotz, J. C. (1977). *Inorganic Chemistry*, p. 545, Saunders, Philadelphia.
Reiff, W. M. (1973). *Coord. Chem. Rev.*, **10**, 37.
Reiff, W. M., Frankel, R. B., Little, B. F. & Long, G. J. (1974). *Inorg. Chem.*, **13**, 2153. *Chem. Phys. Lett.*, **28**, 68.
Robin, M. B. & Day, P. (1967). *Adv. Inorg. Chem. Radiochem.*, **10**, 247.
Sakai, T. & Tominaga, T. (1975). *Bull. Chem. Soc. Japan*, **48**, 3168.
Sams, J. R. & Tsin, T. B. (1979). In: *The Porphyrins*, Vol. IV, ed. D. Dolphin, p. 425, Academic Press, New York.
Shenoy, G. K. (1984). In: *Mössbauer Spectroscopy Applied to Inorganic Chemistry*, Vol. I, ed. G. J. Long, p. 57, Plenum Press, New York.
Sidgwick, N. V. & Bailey, R. W. (1934). *Proc. Roy. Soc.* (London), **144A**, 521.
Sorai, M. & Seki, S. (1974). *J. Phys. Solids*, **35**, 555.
Spiering, H. (1984). In: *Mössbauer Spectroscopy Applied to Inorganic Chemistry*, Vol. I, ed. G. J. Long, p. 77, Plenum Press, New York.
Stevens, E. D., De Lucia, M. L. & Coppens, P. (1980). *Inorg. Chem.*, **19**, 813.
Stevens, E. D. (1981). *J. Amer. Chem. Soc.*, **103**, 5087.
Stevens, E. D. (1982). In: *Electron Distributions and the Chemical Bond*, eds. P. Coppens & M. B. Hall, p. 331, Plenum Press, New York.
Takemoto, J. H. & Hutchinson, B. (1973). *Inorg. Chem.*, **12**, 705.
Walton, E. G., Corvan, P. J., Brown, D. B. & Day, P. (1976). *Inorg. Chem.*, **15**, 1737.
Walton, E. G., Brown, D. B., Wong, H. & Reiff, W. M. (1977). *Inorg. Chem.*, **16**, 2425.
Wei, C. H. & Dahl, L. F. (1968). *J. Amer. Chem. Soc.*, **91**, 3151.
Wright, M. E., Long, G. J., Tharp, D. E. & Nelson, G. O. (1986). *Organometallics*, **5**, 779.
Yamanaka, T. & Takeuchi, Y. (1979). *Phys. Chem. Minerals*, **4**, 149.
Yoshihashi, T. & Sano, H. (1975). *Chem. Phys. Lett.*, **34**, 289.
Zerner, M., Gouterman, M. & Kobayashi (1966). *Theor. Chim. Acta*, **6**, 363.

## Chapter 4

Abragam, A. & Pryce, M. H. L. (1951). *Proc. Roy. Soc.* (London), **205A**, 135.

Baines, J. A., Johnson, C. E. & Thomas, M. F. (1983). *J. Phys. C*, **16**, 3579.

Bell, S. H., Weir, M. P., Dickson, D. P. E., Gibson, J. F., Sharp, G. A. & Peters, T. J. (1984). *Biochim. Biophys. Acta*, **787**, 227.

Boersma, F., Cooper, D. M., de Jonge, W. J. M., Dickson, D. P. E., Johnson, C. E. & Tinus, A. M. C. (1982). *J. Phys. C*, **15**, 4141.

de Jongh, L. J. & Miedema, A. R. (1974). *Adv. Phys.*, **23**, 1.

Eibschütz, M., Davidson, G. R. & Cox, D. E. (1973). *AIP Conf. Proc.*, **18**, 386.

Greenwood, N. N. & Gibb, T. C. (1971). *Mössbauer Spectroscopy*, Chapman and Hall, London.

Gupta, G. P., Dickson, D. P. E. & Johnson, C. E. (1978). *J. Phys. C*, **11**, 215.

Hesse, J. & Rubartsch (1974). *J. Phys. E*, **7**, 526.

Hesse, J., Müller, J. B. & Weichman, B. (1979). *J. Physique*, **40**, C2–161.

Johnson, C. E. (1971). In: *Hyperfine Interactions in Excited Nuclei*, eds. G. Goldring & B. Kalish, p. 803, Gordon & Breach, London.

Johnson, J. A. (1984). PhD Thesis, University of Liverpool.

Keller, H. & Savic, J. M. (1983). *Phys. Rev.*, **B28**, 2638.

Kopcewicz, M., Wagner, H. G. & Gonser, U. (1984). *Sol. Stat. Commun.*, **48**, 531.

Kündig, W. (1967). *Nucl. Instr. Meth.*, **48**, 219.

Lines, M. E. & Eibschütz (1983). *Sol. Stat. Commun.*, **34**, 435.

Marshall, W. & Johnson, C. E. (1962). *Journal de Physique et la Radium*, **23**, 733.

Mørup, S., Dumesic, J. A. & Topsøe, H. (1980). In: *Applications of Mössbauer Spectroscopy*, Vol. II, ed. R. L. Cohen, Chap. 1, Academic Press, New York.

Mørup, S. (1983). *J. Magn. Magn. Mater.*, **37**, 39.

Mørup, S., Madsen, M. B., Franck, J., Villadsen, J. & Koch, C. J. W. (1983). *J. Magn. Magn. Mater.*, **40**, 163.

Prelorendjos, L. A. (1980). PhD Thesis, University of Liverpool.

Rush, J. D., Simopoulos, A., Thomas, M. F. & Wanklyn, B. M. (1976). *Sol. Stat. Commun.*, **18**, 1039.

Tsunato, T. & Murao, T. (1971). *Physica*, **51**, 186.

Window, B. (1971). *J. Phys. E*, **4**, 401.

Window, B. (1974). *J. Phys. E*, **4**, 329.

## Chapter 5

Bauminger, E. R., Froindlich, D., Nowik, I., Ofer, S., Felner, I. & Mayer, I. (1973). *Phys. Rev. Lett.*, **30**, 1053.

Bauminger, E. R., Felner, I., Levron (Lebenbaum), D., Nowik, I. & Ofer, S. (1974). *Phys. Rev. Lett.*, **33**, 890.

Bender, O. & Schroeder, K. (1979). *Phys. Rev.*, **B19**, 3399.

Berkooz, O., Malamud, M. & Shtrikman, S. (1968). *Sol. Stat. Commun.*, **6**, 185.

Chudley, C. T. & Elliott, R. J. (1961). *Proc. Phys. Soc.* (London), **77**, 353.

Dattagupta, S. (1981). *Hyperfine Interactions*, **11**, 77.

Dattagupta, S. (1983). In: *Advances in Mössbauer Spectroscopy, Applications to Physics, Chemistry and Biology*, eds. B. V. Thosar, J. K. Srivastava, P. K. Iyengar & S. C. Bhargava, Elsevier, Amsterdam and New York.

Dattagupta, S., Shenoy, G. K., Dunlap, B. D. & Asch, L. (1977). *Phys. Rev.*, **B16**, 3893.

Heidemann, A., Kaindl, G., Salomon, D., Wipf, H. & Wortmann, G. (1976). *Phys. Rev. Lett.*, **36**, 213.

Lewis, S. J. & Flinn, P. A. (1969). *Appl. Phys. Lett.*, **15**, 331.

Lindley, D. H. & Debrunner, P. G. (1966). *Phys. Rev.*, **146**, 199.

Shenoy, G. K., Dunlap, B. D., Dattagupta, S. & Asch, L. (1976). *Phys. Rev. Lett.*, **37**, 539.
Singwi, K. S. & Sjölander, A. (1960). *Phys. Rev.*, **120**, 1093.
Song, C.-j. & Mullen, J. G. (1976). *Phys. Rev.*, **B14**, 2761.


Chapter 6

Armon, H., Bauminger, E. R., Diamant, A., Nowik, I. & Ofer, S. (1973). *Phys. Lett.*, **44A**, 279.
Bashkirov, S. S. & Selyntin, G. Y. (1968). *Soviet Phys.: Sol. Stat. Phys.*, **9**, 2284; *Phys. Stat. Sol.*, **26**, 253.
Bauminger, E. R., Diamant, A., Felner, I., Nowik, I. & Ofer, S. (1974). *Phys. Lett.*, **50A**, 321.
Bauminger, E. R., Cohen, S. G., Nowik, I., Ofer, S. & Yariv, J. (1983). *Proc. Natl. Acad. Sci. USA*, **80**, 736. *Hyperfine Interactions*, **15/16**, 881.
Bauminger, E. R., Nowik, I., Ofer, S. & Virguin-Heitner, C. (1985). *Polymer*, **26**, 1829.
Bläsius, A., Preston, R. S. & Gonser, U. (1979). *Z. Phys. Chem. Neue Folge*, **115**, 373.
Boyle, A. J. F., Bunbury, D. St. P., Edwards, C. & Hall, H. E. (1961). *Proc. Phys. Soc.* (London), **77**, 129.
Bryukhanov, V. A., Delyagin, N. N. & Kagan, Yu. M. (1963). *Soviet Physics: JETP*, **45**, 1372.
Bunbury, D. St. P., Elliot, S. A., Hall, H. E. & Williams, J. M. (1963). *Phys. Lett.*, **6**, 34.
Chudley, C. T. & Elliot, R. J. (1961). *Proc. Phys. Soc.* (London), **77**, 353.
Cohen, S. G., Bauminger, E. R., Nowik, I. & Ofer, S. (1981). *Phys. Rev. Lett.*, **46**, 1244.
Craig, P. P. & Suttin, N. (1963). *Phys. Rev. Lett.*, **11**, 460.
Dattagupta, S. (1976). *Phys. Rev.*, **B14**, 1329.
Dibar-Ure, M. C. & Flinn, P. A. (1973). *Appl. Phys. Lett.*, **23**, 587.
Dwivedi, A., Pederson, T. & Debrunner, P. G. (1979). *J. Physique*, **40**, C2–531.
Feder, D. & Nowik, I. (1979). *J. Magn. Magn. Mater.*, **12**, 149.
Flinn, P. A. (1980). In: *Applications of Mössbauer Spectroscopy*, Vol. II, ed. R. L. Cohen, p. 393, Academic Press, New York.
Frolov, E. N., Likhtenstein, G. I. & Goldanskii, V. I. (1975). In: *Proceedings VI International Conference on Mössbauer Spectroscopy*, Vol. II, Cracow, eds. A. Z. Hrynkiewicz & J. A. Sawicki, p. 319.
Goldanskii, V. I. & Korytko, L. A. (1973). *JETP Lett.*, **17**, 227.
Goldanskii, V. I., Makarov, E. F. & Khrapov, V. V. (1963). *Phys. Lett.*, **3**, 344.
Heidemann, A., Wipf, H. & Wortmann, G. (1978). *Hyperfine Interactions*, **4**, 844.
Heilmann, I., Olsen, B. & Jensen, J. H. (1974). *J. Phys. C*, **7**, 4355.
Herber, R. H. & Maeda, Y. (1980). *Physica*, **99B**, 352.
Howard, D. G. & Nussbaum, R. H. (1980). *Surf. Sci.*, **93**, L105.
Josephson, B. D. (1960). *Phys. Rev. Lett.*, **4**, 341.
Karyagin, S. V. (1963). *Dokl. Akad. Nauk. SSSR*, **148**, 1102.
Keller, H. & Debrunner, P. G. (1980). *Phys. Rev. Lett.*, **45**, 68.
Kimball, C. W., Taneja, S. P., Weber, L. & Fradin, F. Y. (1974). In: *Mössbauer Effect Methodology*, Vol. IX, ed. I. J. Gruverman, p. 93, Plenum Press, New York.
Kitchens, T. A., Craig, P. P. & Taylor, R. D. (1970). In: *Mössbauer Effect Methodology*, Vol. V, ed. I. J. Gruverman, p. 123, Plenum Press, New York.
Knapp, E. W., Fischer, S. F. & Parak, F. (1983). *J. Chem. Phys.*, **78**, 4701.
La Price, W. J. & Uhrich, D. L. (1979). *Chem. Phys.*, **71**, 1438.
Litterst, F. J., Lerf, A., Nuyken, D. & Alcala, H. (1982). *Hyperfine Interactions*, **12**, 317.
Mannheim, P. D. (1968). *Phys. Rev.*, **165**, 1011.
Mannheim, P. D. & Simopoulos, A. I. (1968). *Phys. Rev.*, **165**, 845.
Mantl, S., Petry, W., Schroeder, K. & Vogl, G. (1983). *Phys. Rev.*, **27**, 5313.

Maradudin, A. A. (1966). In: *Solid State Physics*, Vol. XVIII, p. 274, Academic Press, New York.

Mayo, K. H., Parak, F. & Mössbauer, R. L. (1981). *Phys. Lett.*, **82A**, 468.

Mössbauer, R. L. (1958). *Z. Physik.*, **151**, 124; *Naturwissenschaften*, **45**, 538.

Mössbauer, R. L. (1983). In: *Trends in Mössbauer Spectroscopy – Second Seeheim Workshop*, eds. P. Gütlich & G. M. Kalvius, p. 86, The University of Mainz Press, Mainz.

Mullen, J. G. (1980). *Phys. Lett.*, **79A**, 457.

Mullen, J. G. (1982). In: *Proceedings of the International Conference on the Applications of the Mössbauer Effect*, p. 29, Indian National Science Academy, New Delhi.

Nowik, I., Bauminger, E. R., Cohen, S. G. & Ofer, S. (1985). *Phys. Rev.*, **A31**, 2299.

Nowik, I., Cohen, S. G., Bauminger, E. R. & Ofer, S. (1983). *Phys. Rev. Lett.*, **50**, 152.

Nowik, I. & Felner, I. (1985). *Physica*, **130B**, 433.

Nussbaum, R. H. (1966). In: *Mössbauer Effect Methodology*, Vol. II, ed. I. J. Gruverman, p. 3, Plenum Press, New York.

Ofer, S., Bauminger, E. R., Cohen, S. G. & Nowik, I. (1983). In: *Structure and Function of Iron Storage and Transport Proteins*, eds. I. Urushizaki *et al.*, p. 59.

Ofer, S., Nowik, I., Bauminger, E. R., Papaefthymiou, G. C., Frankel, R. B. & Blakemore, R. P. (1984). *Biophys. J.*, **46**, 57.

Parak, F. & Formanek, H. (1971). *Acta Crystallogr. Sect. A*, **27**, 573.

Parak, F., Frolov, E. N., Mössbauer, R. L. & Goldanskii, V. I. (1981). *J. Mol. Biol.*, **145**, 825.

Parak, F., Knapp, E. W. & Kucheida, D. (1982). *J. Mol. Biol.*, **161**, 177.

Petry, W. (1983). In: *Trends in Mössbauer Spectroscopy – Second Seeheim Workshop*, eds. P. Gütlich & G. M. Kalvius, p. 193, The University of Mainz Press, Mainz.

Petry, W., Vogl, G. & Mansel, W. (1980). *Phys. Rev. Lett.*, **45**, 1862.

Petry, W., Vogl, G. & Mansel, W. (1982). *Z. Phys.*, **B46**, 319.

Pfannes, H. D. & Gonser, U. (1973). *Appl. Phys.*, **1**, 93.

Pound, R. V. & Rebka, G. A. (1959). *Phys. Rev. Lett.*, **3**, 439.

Probst, F., Wagner, F. E. & Karger, M. (1980). *J. Phys. F*, **10**, 2081.

Ruby, S. (1973). In: *Perspectives in Mössbauer Spectroscopy*, eds. S. G. Cohen & M. Pasternak, p. 181, Plenum Press, New York.

Seh, M. & Nowik, I. (1981). *J. Magn. Magn. Mater.*, **22**, 239.

Shenoy, G. K. (1973). In: *Perspectives in Mössbauer Spectroscopy*, eds. S. G. Cohen & M. Pasternak, p. 141, Plenum Press, New York.

Shenoy, G. K. & Friedt, J. M. (1973). *Phys. Rev. Lett.*, **31**, 419.

Simopoulos, A. I., Wickman, H., Kostikas, A. & Petrides, D. (1970). *Chem. Phys. Lett.*, **7**, 615.

Singwi, K. S. & Sjölander, A. (1960). *Phys. Rev.*, **120**, 1093.

Visscher, K. M. (1960). *Ann. Phys. (N.Y.)*, **9**, 194.

Wipf, H. & Heidemann, A. (1980). *J. Phys. C*, **13**, 5757.

Wolf, D. (1977). *Appl. Phys. Lett.*, **30**, 617.

Wordel, R. & Wagner, F. E. (1984). *J. Less. Com. Met.*, **101**, 427.

# Index